CHEMISTRY OF CHEMICAL BONDING

CHEMISTRY OF CHEMICAL BONDING

Jyoti Kumar Roshan

ANMOL PUBLICATIONS PVT. LTD.
NEW DELHI - 110 002 (INDIA)

ANMOL PUBLICATIONS PVT. LTD.
H.O.: 4374/4B, Ansari Road, Daryaganj,
New Delhi-110 002 (India)
Ph.: 23278000, 23261597
B.O.: No. 1015, Ist Main Road, BSK IIIrd Stage
IIIrd Phase, IIIrd Block,
Bangalore - 560 085 (India)
Visit us at: www.anmolpublications.com

Chemistry of Chemical Bonding

ISBN 978-81-261-3674-2

PRINTED IN INDIA

Printed at Mehra Offset Press, Delhi.

Contents

Preface

The beginning student of chemistry must have a knowledge of the theory which forms the basis for our understanding of chemistry and he must acquire this knowledge before he has the mathematical background required for a rigorous course of study in quantum mechanics. The present approach is designed to meet this need by stressing the physical or observable aspects of the theory through an extensive use of the electronic charge density. The manner in which the negative charge of an atom or a molecule is arranged in three-dimensional space is determined by the electronic charge density distribution. Thus, it determines directly the sizes and shapes of molecules, their electrical moments and, indeed, all of their chemical and physical properties.

Since the charge density describes the distribution of negative charge in real space, it is a physically measurable quantity. Consequently, when used as a basis for the discussion of chemistry, the charge density allows for a direct physical picture and interpretation.

In this treatment, the classifications of bonding, ionic or covalent, are retained, but they are given ***physical*** definitions in terms of the actual distribution of charge within the molecule. In covalent bonding the valence charge density is distributed over the whole molecule and the attractive forces responsible for binding the nuclei are exerted by the charge density equally shared between them in the internuclear region. In ionic bonding, the valence charge density is localized in the region of a single nucleus and in this extreme of binding the charge density localized on a single nucleus exerts the attractive force which binds both nuclei.

Author

Preface

Chapter 1

Introduction

Ionic Bonding

In general chemistry you learned that having eight electrons (an octet) in the valence shell of the noble gases (group 8A) gives those elements special stability. We also have seen that the reactions of most elements are driven by their desire to attain a noble gas electronic configuration. The alkali metals in group 1A have a single electron in their valence shell and usually give up this electron in reactions to achieve a noble gas electronic configuration. Unlike the alkali metals which tend to give up their valence electron to achieve a noble gas configuration, the halogens in group 7A (with 7 valence electrons) tend to gain an electron in reactions to attain a full octet of electrons. Elements on the right side of the periodic table (nonmetals) react by gaining electrons and elements on the left side of the periodic table (metals) react by losing electrons. The type of bonding that occurs when a nonmetal accepts an electron from a metal is referred to as ionic bonding and the bond is called an ionic bond. This is the type of bonding that occurs in ionic solids like sodium chloride, NaCl.

Covalent Bonding

In the previous section we saw that elements on the left and right sides of the periodic table form ionic bonds by gaining or losing electrons. It may at first glance seem that elements in the middle of the periodic table would form bonds in a similar manner, but that is not the case. Carbon and other elements in the middle of the periodic table form bonds by sharing electrons in covalent bonds. Carbon forms bonds by sharing electrons because it would be very difficult to either gain or lose the four electrons necessary to achieve a noble gas electronic configuration. (Covalent and ionic bonding are extremes in the bonding picture and most bonds have some characteristics of both ionic bonds and covalent bonds.) These shared electron bonds are easily drawn using electron-dot structures or Lewis structures.

Valence electrons not used for bonding are referred to as lone pair electrons or nonbonding electrons. In Lewis structures all valence electrons including lone pairs are always shown. Notice that each chlorine atom in carbon tetrachloride shown has 3 lone pairs of electrons (nonbonding electrons) associated with it. We will see later that these lone pairs of electrons play a role in determining the chemistry of organo-halogen compounds. We explain the reactions and bonding that occurs in organic molecules by monitoring the valence electrons associated with the atoms. In addition to Lewis structures, molecules can also be drawn using "Kekule" structures (also known as line-bond structures) in which the two-electron covalent bond is shown as a line joining the two atoms. Although nonbonding electrons are usually not shown in "Kekule" structures it is very important to keep track of them since they are the electrons that are involved in organic reactions. A "Kekule" structure of carbon tetrachloride is shown below. Notice that the lone pairs of electrons on the four chlorine atoms are not shown but are understood to be there.

Bonding Theories

Since the majority of the bonds we will encounter in organic chemistry are covalent bonds, we need to look at the possible ways of explaining how covalent bonds are formed. There are two models which explain covalent bonding, Valence Bond Theory and Molecular Orbital Theory. In Valence Bond Theory, bonding is viewed as occurring by the overlap of two atomic orbitals, one from each atom. Each atom's orbital contains a single electron and a bond is formed by the electrons, now paired in overlapping orbitals, holding the two nuclei together. Every covalent bond has a characteristic bond strength and bond length. The bond length is defined as the distance between the two nuclei. Bond strength is defined as the amount of energy needed to homolytically break a bond (each of the nuclei takes one of the two electrons from the bond). A normal carbon-chlorine bond has a bond length of 1.78 Angstroms (1.78×10^{-10} meters) and a bond strength of (339 kJ/mol or 81 kcal/mol).

In Molecular Orbital Theory, bonding is explained in terms of the mathematical combination of atomic orbitals to form molecular orbitals. The newly formed orbitals are called molecular orbitals because they belong to the entire molecule. The combination of two atomic orbitals leads to two molecular orbitals, a bonding molecular orbital and an antibonding molecular orbital.

METALS -NONMETALS

ALKALI METALS

The alkali metals are the first group of the periodic table of the elements. These metals are lithium (Li), sodium (Na), potassium (K), rubidium (Rb), caesium (Cs) and francium (Fr). They have got one outer electron and the connection to the atomic nucleus is very weak. They are the least noble substances. In the air they react very fast with oxygen and so,they are covered in an oxide layer. Alkali metals are kept in liquid hydrocarbons like petroleum. The reactivity grows from lithium to caesium. They react suddenly with water, then they burn or can even explode. Because of the big radius of caesium his outer electron is very free, so caesium reacts so strong with oxygen that it begins to burn in the air. When these metals react with water the product is a caustic solution, like caustic soda solution (NaOH). They are very soft, too.

ALKALINE EARTH METALS

The alkaline earth metals are the second group of the periodic table of the elements. These metals are beryllium (Be), magnesium (Mg), calcium (Ca), strontium (Sr), barium (Ba) and radium (Ra). These elements have got two outer electrons and so they did not react as strong as the alkali metals, because also can make caustic solutions. These elements are much harder than the alkali metals. Magnesium is an element which do not only burns in oxygen, but in all substances which contain oxygen. So magnesium burns in water, too. Divers use magnesium flares to see something under water.

Boron Group

The boron group is the third group of the periodic table of the elements. These elements are boron (B), aluminium (Al), gallium (Ga), Indium (In) and thallium (Tl). Boron is a metalloid and there other four elements are metals. The most used element of this group is aluminium. They elements are reactively. When aluminium is used it is covered in an oxide layer.

CARBON SILICON GROUP

The carbon silicon group is the fourth group of the periodic table of the elements. These elements are carbon (C), silicon (Si), germanium (Ge), tin (Sn) and lead (Pb). Carbon is so noble that it is a nonmetal. Silicon and germanium are metalloids and tin and lead are metals. Lead is a very soft metal. Silicon is very often found in the compound silicon dioxide (quartz). Silicon and germanium are very important semiconductor and so they are very important for elektronics.

PNICTOGENS

The pnictogens are the fifth group of the periodic table of the elements. These elements are nitrogen (N), phosphorus (P), arsenic (As), antimony (Sb) and bismuth (Bi). Nitrogen and phosphorus are nonmetals. Arsenic and antimony are metalloids and bismuth is a metal. 80% of the earth atmosphere are nitrogen, but in molecules (N_2).

CHALCOGENS

The chalcogens are the sixth group of the periodic table of the elements. These elements are oxygen (O), sulfur (S), selenium (Se), tellurium (Te) and polonium (Po). Oxygen and sulfur and nonmetals. Selenium and tellurium are metalloids and polonium is a metal.

Halogens

The halogens are the seventh group of the periodic table of the elements. These elements are fluorine (F), chlorine (Cl), bromine (Br), iodine (I) and astatine (At). Fluorine, chlorine, bromine and iodine are nonmetals. Astatine is a metalloid. The halogens only need one electron to reach the noble gas configuration and so they are very reactively. Fluor is the most reactively element of all elements. Many substances (for example wood) begin to burn, because of the contact with fluorine. Fluorine is a colourless gas. The other halogens are very dangerous, because they are very reactively and very corrosive, too. Clorine was used as a strong corrosive poison in the First World War. It is yellow-green. Bromine is a dark liquid at room temperature and like all other halogens a very corrosive substance.

NOBLE GASES

The noble gases are the eigthth group of the periodic table of the elements. These elements are helium (He), neon (Ne), argon (Ar), krypton (Kr), xenon (Xe) and radon (Rn). They are all nonmetals and like the name it says they are gases. They have got 8 outer electron (helium: 2 outer electrons) and they nearly do not react at all. They are the most noble elements.

TRANSITION METALS

We have not got mentioned the most elements of the periodic table of the elements. These are the transition metals. They metals where the other atomic shells are filled. Famous metals are for example gold (Au), silver (Ag), iron (Fe), copper (Cu), mercury (Hg) and many others. The lanthanoids (rare-earths) and the actionoids belong also to the transition metals. Famous actinoids are the both radioactive elements uranium (U) and plutonium (Pu).

Hydrogen

Hydrogen is a chemical element represented by the symbol H and an atomic number of 1. At standard temperature and pressure it is a colorless, odorless, nonmetallic, tasteless, highly flammable diatomic gas (H_2). With an atomic mass of 1.00794 g/mol, hydrogen is the lightest element.

Hydrogen is the most abundant of the chemical elements, constituting roughly 75% of the universe's elemental mass. Stars in the main sequence are mainly composed of hydrogen in its plasma state. Elemental hydrogen is relatively rare on Earth, and is industrially produced from hydrocarbons such as methane, after which most elemental hydrogen is used "captively" (meaning locally at the production site), with the largest markets about equally divided between fossil fuel upgrading (e.g., hydrocracking) and in ammonia production (mostly for the fertilizer market). Hydrogen may be produced from water using the process of electrolysis, but this process is presently significantly more expensive commercially than hydrogen production from natural gas.

The most common naturally occurring isotope of hydrogen, known as protium, has a single proton and no neutrons. In ionic compounds it can take on either a positive charge (becoming a cation composed of a bare proton) or a negative charge (becoming an anion known as a hydride). Hydrogen can form compounds with most elements and is present in water and most organic compounds. It plays a particularly important role in acid-base chemistry, in which many reactions involve the exchange of protons between soluble molecules. As the only neutral atom for which the Schrödinger equation can be solved analytically, study of the energetics and bonding of the hydrogen atom has played a key role in the development of quantum mechanics.

NOMENCLATURE

Hydrogen, Latin: 'hydrogenium', is from Ancient Greek (*hydor*): "water" and (*genes*): "forming". Ancient Greek (*geinomai*): "to beget or sire")

The word "hydrogen" has several different meanings;

1. The *name of an element.*
2. An *atom,* sometimes called "H dot", that is abundant in space but essentially absent on Earth, because it dimerizes.
3. A *diatomic molecule* that occurs naturally in trace amounts in the Earth's atmosphere; chemists increasingly refer to H_2 as *dihydrogen,*or *hydrogen molecule,* to distinguish this molecule from atomic hydrogen and hydrogen found in other compounds.
4. The atomic *constituent* within all organic compounds, water, and many other chemical compounds.

The *elemental* forms of hydrogen should not be confused with hydrogen as it appears in chemical compounds.

HISTORY

Discovery of H_2

Hydrogen gas, H_2, was first artificially produced and formally described by T. Von Hohenheim via the mixing of metals with strong acids. He was unaware that the flammable gas produced by this chemical reaction was a new chemical element. In 1671, Robert Boyle rediscovered and described the reaction between iron filings and dilute acids, which results in the production of hydrogen gas.In 1766, Henry Cavendish was the first to recognize hydrogen gas as a discrete substance, by identifying the gas from a metal-acid reaction as "inflammable air" and further finding that the gas produces water when burned. Cavendish had stumbled on hydrogen when experimenting with acids and mercury. Although he wrongly assumed that hydrogen was a liberated component of the mercury rather than the acid, he was still able to accurately describe several key properties of hydrogen. He is usually given credit for its discovery as an element. In 1783, Antoine Lavoisier gave the element the name of hydrogen when he (with Laplace) reproduced Cavendish's finding that water is produced when hydrogen is burned. Lavoisier's name for the gas won out.

One of the first uses of H_2 was for balloons, and later airships. The H_2 was obtained by reacting sulphuric acid and metallic iron. Infamously, H_2 was used in the Hindenburg airship that was destroyed in a midair fire. The highly flammable hydrogen (H_2) was later replaced for airships and most balloons by the unreactive helium (He).

Role in History of Quantum Theory

Because of its relatively simple atomic structure, consisting only of a proton and an electron, the hydrogen atom, together with the spectrum of light produced from it or absorbed by it, has been central to the development of the theory of atomic structure. Furthermore, the corresponding simplicity of the hydrogen molecule and the corresponding cation H_2^+ allowed fuller understanding of the nature of the chemical bond, which followed shortly after the quantum mechanical treatment of the hydrogen atom had been developed in the mid-1920s.

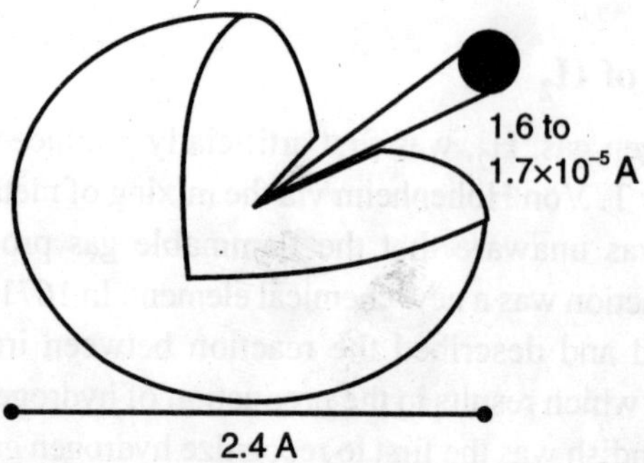

One of the first quantum effects to be explicitly noticed (but not understood at the time) was a Maxwell observation involving hydrogen, half a century before full quantum mechanical theory arrived. Maxwell observed that the specific heat capacity of H_2 unaccountably departs from that of a diatomic gas below room temperature and begins to increasingly resemble that of a monatomic gas at cryogenic temperatures. According to quantum theory, this behaviour arises from the spacing of the (quantized) rotational energy levels, which are particularly wide-spaced in H_2 because of its low mass. These widely spaced levels inhibit equal partition of heat energy into rotational motion in hydrogen at low temperatures. Diatomic gases composed of heavier atoms do not have such widely spaced levels and do not exhibit the same effect.

Natural Occurrence

A giant region of ionized hydrogen in the Triangulum Galaxy.Hydrogen is the most abundant element in the universe, making up 75% of normal matter by mass and over 90% by number of atoms. This element is found in great abundance in stars and gas giant planets. Molecular clouds of H_2 are associated with star formation. Hydrogen plays a vital role in powering stars through proton-proton reaction nuclear fusion.

Throughout the universe, hydrogen is mostly found in the atomic and plasma states whose properties are quite different from molecular hydrogen. As a plasma, hydrogen's electron and proton are not bound together, resulting in very high electrical conductivity and high emissivity (producing the light from the sun and other stars). The charged particles are highly influenced by magnetic and electric fields. For example, in the solar wind they interact with the Earth's magnetosphere giving rise to Birkeland currents and the aurora. Hydrogen is found in the neutral atomic state in the Interstellar medium. The large amount of neutral hydrogen found in the damped Lyman-alpha systems is thought to dominate the cosmological baryonic density of the Universe up to redshift z=4.

Under ordinary conditions on Earth, elemental hydrogen exists as the diatomic gas, H_2. However, hydrogen gas is very rare in the Earth's atmosphere (1 ppm by volume) because of its light weight, which enables it to escape from Earth's gravity more easily than heavier gases. Although H atoms and H_2 molecules are abundant in interstellar space, they are difficult to generate, concentrate, and purify on Earth. Still, hydrogen is the third most abundant element on the Earth's surface.Most of the Earth's hydrogen is in the form of chemical compounds such as hydrocarbons and water. Hydrogen gas is produced by some bacteria and algae and is a natural component of flatus. Methane is a hydrogen source of increasing importance.

The Hydrogen Atom

Electron Energy Levels

Depiction of a hydrogen atom showing the diameter as about twice the Bohr model radius.(Image not to scale)

The ground state energy level of the electron in a hydrogen atom is 13.6 eV, which is equivalent to an ultraviolet photon of roughly 92 nm.

The energy levels of hydrogen can be calculated fairly accurately using the Bohr model of the atom, which conceptualizes the electron as "orbiting" the proton in analogy to the Earth's orbit of the sun. However, the electromagnetic force attracts electrons and protons to one another, while planets and celestial objects are attracted to each other by gravity. Because of the discretization of angular momentum postulated in early quantum mechanics by Bohr, the electron in the Bohr model can only occupy certain allowed distances from the proton, and therefore only certain allowed energies. A more accurate description of the hydrogen atom comes from a purely quantum mechanical treatment that uses the Schrödinger equation or the equivalent Feynman path integral formulation to calculate the probability density of the electron around the proton. Treating the electron as a matter wave reproduces chemical results such as shape of the hydrogen atom more naturally than the particle-based Bohr model, although the energy and spectral results are the same. Modeling the system fully using the reduced mass of nucleus and electron (as one would do in the two-body problem in celestial mechanics) yields an even better formula for the hydrogen spectra, and also the correct spectral shifts for the isotopes deuterium and tritium. Very small adjustments in energy levels in the hydrogen atom, which correspond to actual spectral effects, may be determined by using a full quantum mechanical theory which corrects for the effects of special relativity, and by accounting for quantum effects arising from production of virtual particles in the vacuum and as a result of electric fields.

In hydrogen gas, the electronic ground state energy level is split into hyperfine structure levels because of magnetic effects of the quantum mechanical spin of the electron and proton. The energy of the atom when the proton and electron spins are aligned is higher than when they are not aligned. The transition between these two states can occur through emission of a photon through a magnetic dipole transition. Radio telescopes can detect the radiation produced in this process, which is used to map the distribution of hydrogen in the galaxy.

Isotopes

Protium, the most common isotope of hydrogen, has one proton and one electron. Unique among all stable isotopes, it has no neutrons.

Hydrogen has three naturally occurring isotopes, denoted ^1H, ^2H, and ^3H. Other, highly unstable nuclei (^4H to ^7H) have been synthesized in the laboratory but not observed in nature.

- ^{1}H is the most common hydrogen isotope with an abundance of more than 99.98%. Because the nucleus of this isotope consists of only a single proton, it is given the descriptive but rarely used formal name *protium*.
- ^{2}H, the other stable hydrogen isotope, is known as *deuterium* and contains one proton and one neutron in its nucleus. Deuterium comprises 0.0026 – 0.0184% (by mole-fraction or atom-fraction) of hydrogen samples on Earth, with the lower number tending to be found in samples of hydrogen gas and the higher enrichments (0.015% or 150 ppm) typical of ocean water. Deuterium is not radioactive, and does not represent a significant toxicity hazard. Water enriched in molecules that include deuterium instead of normal hydrogen is called heavy water. Deuterium and its compounds are used as a non-radioactive label in chemical experiments and in solvents for ^{1}H-NMR spectroscopy. Heavy water is used as a neutron moderator and coolant for nuclear reactors. Deuterium is also a potential fuel for commercial nuclear fusion.
- ^{3}H is known as *tritium* and contains one proton and two neutrons in its nucleus. It is radioactive, decaying into Helium-3 through beta decay with a half-life of 12.32 years. Small amounts of tritium occur naturally because of the interaction of cosmic rays with atmospheric gases; tritium has also been released during nuclear weapons tests. It is used in nuclear fusion reactions, as a tracer in isotope geochemistry, and specialized in self-powered lighting devices. Tritium was once routinely used in chemical and biological labeling experiments as a radiolabel.

Hydrogen is the only element that has different names for its isotopes in common use today. The symbols D and T (instead of ^{2}H and ^{3}H) are sometimes used for deuterium and tritium, but the corresponding symbol P is already in use for phosphorus and thus is not available for protium. IUPAC states that while this use is common it is not preferred.

Elemental Molecular Forms

First tracks observed in liquid hydrogen bubble chamber at the Bevatron.

There are two different types of diatomic hydrogen molecules that differ by the relative spin of their nuclei.In the orthohydrogen form, the spins of the two protons are parallel and form a triplet state; in the parahydrogen form the spins are antiparallel and form a singlet. At standard temperature and pressure, hydrogen gas contains about 25% of the para form and 75% of the ortho form, also known as the "normal form". The equilibrium ratio of orthohydrogen to parahydrogen depends on temperature, but since the ortho form is an excited state and has a higher energy than the para form, it is unstable and cannot be purified. At very low temperatures, the equilibrium state is composed almost exclusively of the para form. The physical properties of pure parahydrogen differ slightly from those of the normal form.The ortho/para distinction also occurs in other hydrogen-containing molecules or functional groups, such as water and methylene.

The uncatalyzed interconversion between para and ortho H_2 increases with increasing temperature; thus rapidly condensed H_2 contains large quantities of the high-energy ortho form that convert to the para form very slowly.The ortho/para ratio in condensed H_2 is an important consideration in the preparation and storage of liquid hydrogen: the conversion from ortho to para is exothermic and produces enough heat to evaporate the hydrogen liquid, leading to loss of the liquefied material. Catalysts for the ortho-para interconversion, such as iron compounds, are used during hydrogen cooling.

A molecular form called protonated molecular hydrogen, or H_3^+, is found in the interstellar medium (ISM), where it is generated by ionization of molecular Hydrogen from cosmic rays. It has also been observed in the upper atmosphere of the planet Jupiter. This molecule is relatively stable in the environment of outer space due to the low temperature and density. H_3^+ is one of the most abundant ions in the Universe, and it plays a notable role in the chemistry of the interstellar medium.

Chemical and Physical Properties

The solubility and adsorption characteristics of hydrogen with various metals are very important in metallurgy and in developing safe ways to store it for use as a fuel. Hydrogen is highly soluble in many compounds composed of rare earth metals and transition metals and can be dissolved in both crystalline and amorphous metals. Hydrogen solubility in metals is influenced by local distortions or impurities in the metal crystal lattice.

Combustion

Hydrogen can combust rapidly in air. It burned rapidly in the *Hindenburg* disaster. Hydrogen gas is highly flammable and will burn at concentrations as low as 4% H_2 in air. The enthalpy of combustion for hydrogen is – 286 kJ/mol; it combusts according to the following balanced equation.

$$2\ H_2(g) + O_2(g) \rightarrow 2\ H_2O(l) + 572\ \text{kJ/mol}$$

When mixed with oxygen across a wide range of proportions, hydrogen explodes upon ignition. Hydrogen burns violently in air. Pure hydrogen-oxygen flames are nearly invisible to the naked eye, as illustrated by the faintness of flame from the main Space Shuttle engines. Thus it is difficult to visually detect if a hydrogen leak is burning. The Hindenburg zeppelin flames seen in the adjacent picture are hydrogen flames colored with material from the covering skin of the zeppelin which contained carbon and pyrophoric aluminium powder, as well as other combustible materials. (Regardless of the cause of this fire, this was clearly primarily a hydrogen fire since skin of the Zeppelin alone would have taken many hours to burn). Another characteristic of hydrogen fires is that the flames tend to ascend rapidly with the gas in air, as illustrated by the Hindenberg flames, causing less damage than hydrocarbon fires. For example, two-thirds of the Hindenburg passengers survived the hydrogen fire, and many of the deaths which occurred were from falling or from gasoline burns.

H_2 reacts directly with other oxidizing elements. A violent and spontaneous reaction can occur at room temperature with chlorine and fluorine, forming the corresponding hydrogen halides: hydrogen chloride and hydrogen fluoride.

Compounds

Covalent and Organic Compounds

While H_2 is not very reactive under standard conditions, it does form compounds with most elements. Millions of hydrocarbons are known, but they are not formed by the direct reaction of elementary hydrogen and carbon (although synthesis gas production followed by the Fischer-Tropsch process to make hydrocarbons comes close to being an exception, as this begins with coal and the elemental hydrogen is generated in situ). Hydrogen can form compounds with elements that are more electronegative, such as halogens (e.g., F, Cl, Br, I) and chalcogens (O, S, Se); in these compounds hydrogen takes on a partial positive charge.

When bonded to fluorine, oxygen, or nitrogen, hydrogen can participate in a form of strong noncovalent bonding called hydrogen bonding, which is critical to the stability of many biological molecules. Hydrogen also forms compounds with less electronegative elements, such as the metals and metalloids, in which it takes on a partial negative charge. These compounds are often known as hydrides.

Hydrogen forms a vast array of compounds with carbon. Because of their general association with living things, these compounds came to be called organic compounds; the study of their properties is known as organic chemistry and their study in the context of living organisms is known as biochemistry. By some definitions, "organic" compounds are only required to contain carbon (as a classic historical example, urea). However, most of them also contain hydrogen, and since it is the carbon-hydrogen bond which gives this class of compounds most of its particular chemical characteristics, carbon-hydrogen bonds are required in some definitions of the word "organic" in chemistry. (This latter definition is not perfect, however, as in this definition urea would *not* be included as an organic compound).

In inorganic chemistry, hydrides can also serve as bridging ligands that link two metal centers in a coordination complex. This function is particularly common in group 13 elements, especially in boranes (boron hydrides) and aluminum complexes, as well as in clustered carboranes.

Hydrides

Compounds of hydrogen are often called hydrides, a term that is used fairly loosely. To chemists, the term "hydride" usually implies that the H atom has acquired a negative or anionic character, denoted H^-. The existence of the hydride anion, suggested by G.N. Lewis for group I and II salt-like hydrides, was demonstrated by Moers with the electrolysis of molten lithium hydride (LiH), that produced a stoichiometric quantity of hydrogen at the anode. For hydrides other than group I and II metals, the term is quite misleading, considering the low electronegativity of hydrogen. An exception in group II hydrides is BeH_2, which is polymeric. In lithium aluminum hydride, the AlH_4^- anion carries hydridic centers firmly attached to the Al(III). Although hydrides can be formed with almost all main-group elements, the number and combination of possible compounds varies widely; for example, there are over 100 binary borane hydrides known, but only one binary aluminum hydride.Binary indium hydride has not yet been identified, although larger complexes exist.

"Protons" and Acids

Oxidation of H_2 formally gives the proton, H^+. This species is central to discussion of acids, though the term proton is used loosely to refer to positively charged or cationic hydrogen, denoted H^+. A bare proton H^+ cannot exist in solution because of its strong tendency to attach itself to atoms or molecules with electrons. To avoid the convenient fiction of the naked "solvated proton" in solution, acidic aqueous solutions are sometimes considered to contain the hydronium ion (H_3O^+) organized into clusters to form $H_9O_4^+$. Other oxonium ions are found when water is in solution with other solvents.

Although exotic on earth, one of the most common ions in the universe is the H_3^+ ion, known as protonated molecular hydrogen or the triatomic hydrogen cation.

Production

H_2 is produced in chemistry and biology laboratories, often as a by-product of other reactions; in industry for the hydrogenation of unsaturated substrates; and in nature as a means of expelling reducing equivalents in biochemical reactions.

Laboratory Syntheses

In the laboratory, H_2 is usually prepared by the reaction of acids on metals such as zinc.

$$Zn + 2\,H^+ \rightarrow Zn^{2+} + H_2$$

Aluminum produces H_2 upon treatment with acids but also with base:

$$2\,Al + 6\,H_2O \rightarrow 2\,Al(OH)_3 + 3\,H_2$$

The electrolysis of water is a simple method of producing hydrogen, although the resulting hydrogen necessarily has less energy content than was required to produce it. A low voltage current is run through the water, and gaseous oxygen forms at the anode while gaseous hydrogen forms at the cathode. Typically the cathode is made from platinum or another inert metal when producing hydrogen for storage. If, however, the gas is to be burnt on site, oxygen is desirable to assist the combustion, and so both electrodes would be made from inert metals. (Iron, for instance, would oxidize, and thus decrease the amount of oxygen given off.) The theoretical maximum efficiency (electricity used vs. energetic value of hydrogen produced) is between 80 – 94%. Bellona Report on Hydrogen

$$2H_2O(aq) \rightarrow 2H_2(g) + O_2(g)$$

In 2007, it was discovered that an alloy of aluminium and gallium in pellet form added to water could be used to generate hydrogen.The process creates also creates alumina, but the expensive gallium, which prevents to formation of an oxide skin on the pellets, can be re-used. This potentially has important implications for a hydrogen economy, since hydrogen can be produced on-site and does not need to be transported.

Industrial Syntheses

Hydrogen can be prepared in several different ways but the economically most important processes involve removal of hydrogen from hydrocarbons. Commercial bulk hydrogen is usually produced by the steam reforming of natural gas.At high temperatures (700 – 1100 °C; 1,300 – 2,000 °F), steam (water vapour) reacts with methane to yield carbon monoxide and H_2.

$$CH_4 + H_2O \rightarrow CO + 3\ H_2$$

This reaction is favored at low pressures but is nonetheless conducted at high pressures (20 atm; 600 inHg) since high pressure H_2 is the most marketable product. The product mixture is known as "synthesis gas" because it is often used directly for the production of methanol and related compounds. Hydrocarbons other than methane can be used to produce synthesis gas with varying product ratios. One of the many complications to this highly optimized technology is the formation of coke or carbon:

$$CH_4 \rightarrow C + 2\ H_2$$

Consequently, steam reforming typically employs an excess of H_2O.

Additional hydrogen from steam reforming can be recovered from the carbon monoxide through the water gas shift reaction, especially with an iron oxide catalyst. This reaction is also a common industrial source of carbon dioxide:

$$CO + H_2O \rightarrow CO_2 + H_2$$

Other important methods for H_2 production include partial oxidation of hydrocarbons:

$$CH_4 + 0.5\ O_2 \rightarrow CO + 2\ H_2$$

and the coal reaction, which can serve as a prelude to the shift reaction above:

$$C + H_2O \rightarrow CO + H_2$$

Hydrogen is sometimes produced and consumed in the same industrial process, without being separated. In the Haber process for the production of ammonia (the world's fifth most produced industrial compound), hydrogen is generated from natural gas.

Hydrogen is also produced in usable quantities as a co-product of the major petrochemical processes of steam cracking and reforming. Electrolysis of brine to yield chlorine also produces hydrogen as a co-product.

Biological Syntheses

H_2 is a product of some types of anaerobic metabolism and is produced by several microorganisms, usually via reactions catalyzed by iron- or nickel-containing enzymes called hydrogenases. These enzymes catalyze the reversible redox reaction between H_2 and its component two protons and two electrons. Evolution of hydrogen gas occurs in the transfer of reducing equivalents produced during pyruvate fermentation to water.

Water splitting, in which water is decomposed into its component protons, electrons, and oxygen, occurs in the light reactions in all photosynthetic organisms. Some such organisms — including the alga *Chlamydomonas reinhardtii* and cyanobacteria — have evolved a second step in the dark reactions in which protons and electrons are reduced to form H_2 gas by specialized hydrogenases in the chloroplast.Efforts have been undertaken to genetically modify cyanobacterial hydrogenases to efficiently synthesize H_2 gas even in the presence of oxygen.

Other rarer but mechanistically interesting routes to H_2 production also exist in nature. Nitrogenase produces approximately one equivalent of H_2 for each equivalent of N_2 reduced to ammonia. Some phosphatases reduce phosphite to H_2.

Applications

Large quantities of H_2 are needed in the petroleum and chemical industries. The largest application of H_2 is for the processing ("upgrading") of fossil fuels, and in the production of ammonia. The key consumers of H_2 in the petrochemical plant include hydrodealkylation, hydrodesulfurization, and hydrocracking.H_2 has several other important uses. H_2 is used as a hydrogenating agent, particularly in increasing the level of saturation of unsaturated fats and oils (found in items such as margarine), and in the production of methanol. It is similarly the source of hydrogen in the manufacture of hydrochloric acid. H_2 is also used as a reducing agent of metallic ores.

Apart from its use as a reactant, H_2 has wide applications in physics and engineering. It is used as a shielding gas in welding methods such as atomic hydrogen welding. H_2 is used as the rotor coolant in electrical generators at power stations, because it has the highest thermal conductivity of any gas. Liquid H_2 is used in cryogenic research, including superconductivity studies. Since H_2 is lighter than air, having a little more than 1/15th of the density of air, it was once widely used as a lifting agent in balloons and airships. However, this use was curtailed after the Hindenburg disaster convinced the public that the gas was too dangerous for this purpose. Hydrogen is still regularly used for the inflation of weather balloons.

Hydrogen's rarer isotopes also each have specific applications. Deuterium (hydrogen-2) is used in nuclear fission applications as a moderator to slow neutrons, and in nuclear fusion reactions. Deuterium compounds have applications in chemistry and biology in studies of reaction isotope effects. Tritium (hydrogen-3), produced in nuclear reactors, is used in the production of hydrogen bombs, as an isotopic label in the biosciences, and as a radiation source in luminous paints.

The triple point temperature of equilibrium hydrogen is a defining fixed point on the ITS-90 temperature scale.

Hydrogen as an Energy Carrier

Hydrogen is not an energy source, except in the hypothetical context of commercial nuclear fusion power plants using deuterium or tritium, a technology presently far from development. The sun's energy comes from nuclear fusion of hydrogen but this process is difficult to achieve on earth. Elemental hydrogen from solar, biological, or electrical sources costs more in energy to make than is obtained by burning it. Hydrogen may be obtained from fossil sources (such as methane) for less energy than required to make it, but these sources are unsustainable, and are also themselves direct energy sources (and are rightly regarded as the basic source of the energy in the hydrogen obtained from them).

Molecular hydrogen has been widely discussed in the context of energy, as a possible carrier of energy on an economy-wide scale. A theoretical advantage of using H_2 as an energy carrier is the localization and concentration of environmentally unwelcome aspects of hydrogen manufacture from fossil fuel energy sources. For example, CO_2 sequestration followed by carbon capture and storage could be conducted at the point of H_2 production from methane. Hydrogen used in transportation would burn cleanly, without carbon emissions. However, the infrastructure costs associated with full conversion to a hydrogen economy would be substantial. In addition, the energy density of both liquid hydrogen and hydrogen gas at any practicable pressure is significantly less than that of traditional fuel sources.

GROUP 1 (ALKALI METALS)

Some atomic and physical properties of the Group 1 elements - lithium, sodium, potassium, rubidium and caesium. You will find separate sections below covering the trends in atomic radius, first ionisation energy, electronegativity, melting and boiling points, and density.The same ideas tend to recur throughout the atomic properties.

Atomic Radius

Atomic Radii of the Group 1 elements

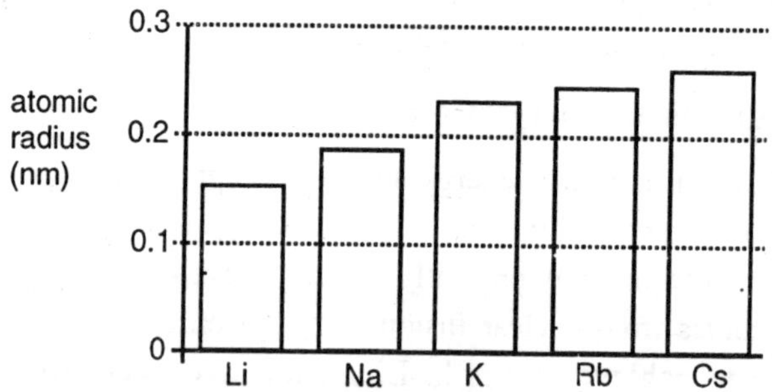

The atomic radius increases as you go down the Group.The radius of an atom is governed by

a. The number of layers of electrons around the nucleus

b. The pull the outer electrons feel from the nucleus.

Compare lithium and sodium:

Li $1s^2 2s^1$

Na $1s^2 2s^2 2p^6 3s^1$

In each case, the outer electron feels a net pull of 1+ from the nucleus. The positive charge on the nucleus is cut down by the negativeness of the inner electrons.

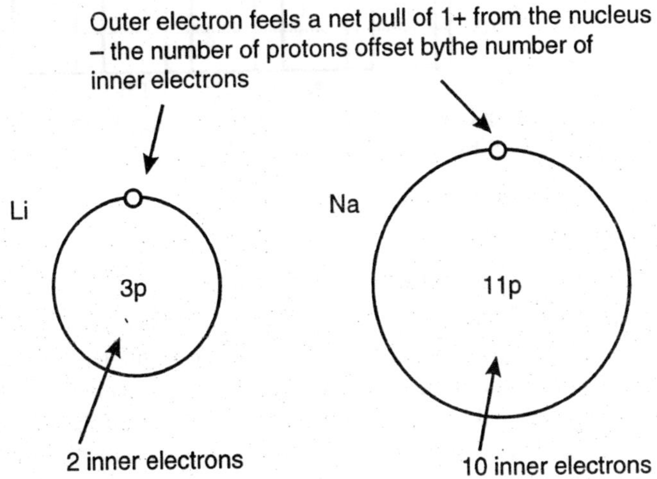

This is equally true for all the other atoms in Group 1. Work it out for potassium if you aren't convinced.The only factor which is going to affect the size of the atom is therefore the number of layers of inner electrons which have to be fitted in around the atom. Obviously, the more layers of electrons you have, the more space they will take up - electrons repel each other. That means that the atoms are bound to get bigger as you go down the Group.

First Ionisation Energy

First ionisation energy is the energy needed to remove the most loosely held electron from each of one mole of gaseous atoms to make one mole of singly charged gaseous ions - in other words, for 1 mole of this process:

$$X_{(g)} \longrightarrow X^{+}_{(g)} + e^{-}$$

First ionisation Energy of the Group 1 elements

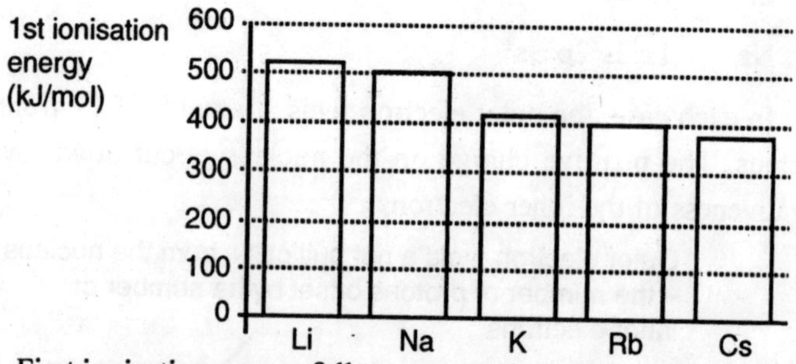

First ionisation energy falls as you go down the group.Ionisation energy is governed by

1. The charge on the nucleus,
2. The amount of screening by the inner electrons,
3. The distance between the outer electrons and the nucleus.

As you go down the Group, the increase in nuclear charge is exactly offset by the increase in the number of inner electrons. Just as when we were talking about atomic radius in each of the elements in this Group, the outer electrons feel a net attraction of 1+ from the centre.However, as you go down the Group, the distance between the nucleus and the outer electrons increases and so they become easier to remove - the ionisation energy falls.

ELECTRONEGATIVITY

Electronegativities of the Group 1elements

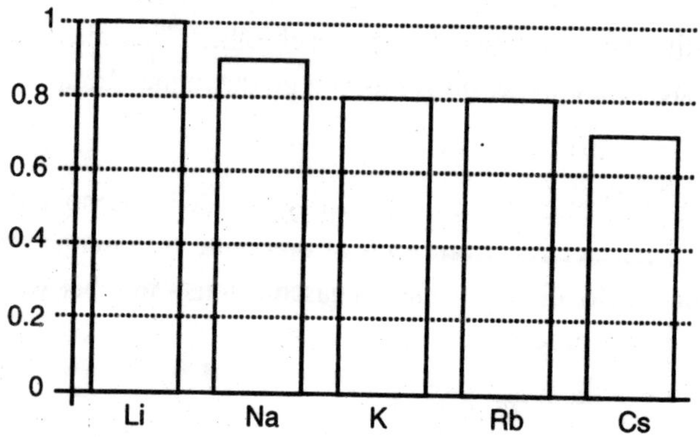

Electronegativity is a measure of the tendency of an atom to attract a bonding pair of electrons. It is usually measured on the Pauling scale, on which the most electronegative element (fluorine) is given an electronegativity of 4.0.

All of these elements have a very low electronegativity.Remember that the most electronegative element, fluorine, has an electronegativity of 4.0.Notice that electronegativity falls as you go down the Group. The atoms become less and less good at attracting bonding pairs of electrons.Imagine a bond between a sodium atom and a chlorine atom. Think of it to start with as a covalent bond - a pair of shared electrons. The electron pair will be dragged towards the chlorine because there is a much greater net pull from the chlorine nucleus than from the sodium one.

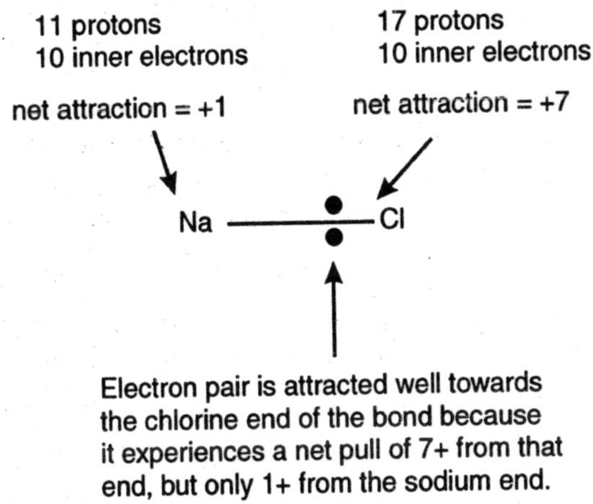

The electron pair ends up so close to the chlorine that there is essentially a transfer of an electron to the chlorine - ions are formed.The large pull from the chlorine nucleus is why chlorine is much more electronegative than sodium is.Now compare this with the lithium-chlorine bond.The net pull from each end of the bond is the same as before, but you have to remember that the lithium atom is smaller than a sodium atom. That means that the electron pair is going to be closer to the net 1+ charge from the lithium end, and so more strongly attracted to it.

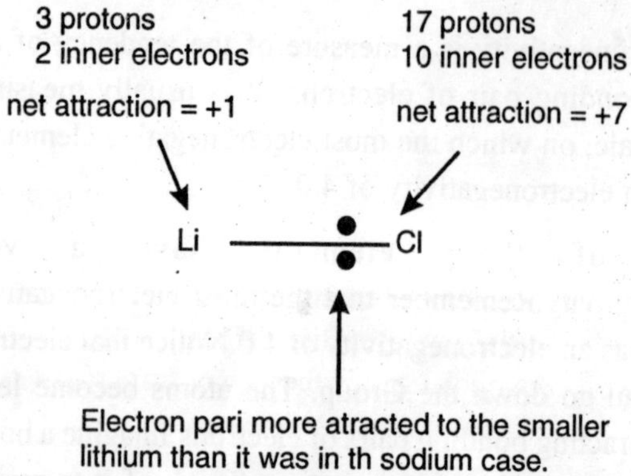

In some lithium compounds there is often a degree of covalent bonding that isn't there in the rest of the Group. Lithium iodide, for example, will dissolve in organic solvents - a typical property of covalent compounds. The iodine atom is so large that the pull from the iodine nucleus on the pair of electrons is relatively weak, and so a fully ionic bond isn't formed.

As the metal atoms get bigger, any bonding pair gets further and further away from the metal nucleus, and so is less strongly attracted towards it. In other words, as you go down the Group, the elements become less electronegative.With the exception of some lithium compounds, these elements all form compounds which we consider as being fully ionic. They are so weakly electronegative that we assume that the electron pair is pulled so far away towards the chlorine (or whatever) that ions are formed.

Melting and Boiling Points

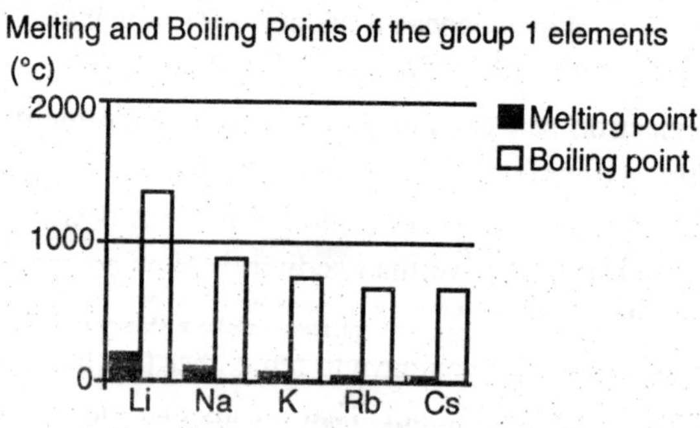

You will see that both the melting points and boiling points fall as you go down the Group.

MELTING AND BOILING POINTS

When you melt any of these metals, the metallic bond is weakened enough for the atoms to move around, and is then broken completely when you boil the metal.

The fall in melting and boiling points reflects the fall in the strength of the metallic bond.The atoms in a metal are held together by the attraction of the nuclei to the delocalised electrons. As the atoms get bigger, the nuclei get further away from these delocalised electrons, and so the attractions fall. That means that the atoms are more easily pulled apart to make a liquid and finally a gas.In the same way that we have already discussed, each of these atoms has a net pull from the nuclei of 1+. The increased charge on the nucleus as you go down the Group is offset by additional levels of screening electrons. All that matters is the distance between the nucleus and the bonding electrons.

Density

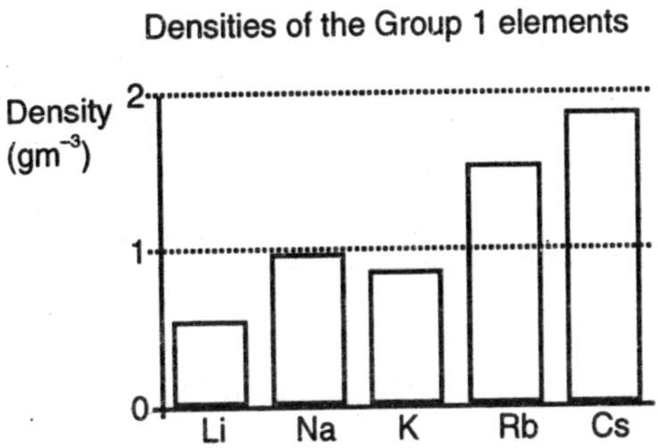

Notice that these are all light metals - and that the first three in the Group are less dense than water (less than 1 g cm^{-3}). That means that the first three will float on water, while the other two sink.The density tends to increase as you go down the Group (apart from the fluctuation at potassium).It is quite difficult to come up with a simple explanation for this, because the density depends on two factors, both of which are changing as you go down the Group.All of these metals have their atoms packed in the same way, so all you have to consider is how many atoms you can pack in a given volume, and what the mass of the individual atoms is. How many you can pack depends, of course, on their volume - and their volume, in turn, depends on their atomic radius.As you go down the Group, the atomic radius increases, and so the volume of the atoms increases as well. That means that you can't pack as many sodium atoms into a given volume as you can lithium atoms.However, as you go down the Group, the mass of the atoms increases. That means that a particular number of sodium atoms will weigh more than the same number of lithium atoms.So 1 cm^3 of sodium will contain fewer atoms than the same volume of lithium, but each atom will weigh more.

REACTIONS OF THE GROUP 1 ELEMENTS WITH WATER

The reactions of the Group 1 elements - lithium, sodium, potassium, rubidium and caesium - with water. It uses these reactions to explore the trend in reactivity in Group 1.All of these metals react vigorously or even explosively with cold water. In each case, a solution of the metal hydroxide is produced together with hydrogen gas.

$$2X_{(s)} + 2H_2O_{(l)} \longrightarrow 2XOH_{(aq)} + H_{2(g)}$$

This equation applies to any of these metals and water - just replace the X by the symbol you want.In each of the following descriptions, I am assuming a very small bit of the metal is dropped into water in a fairly large container.

Lithium

Lithium's density is only about half that of water so it floats on the surface, gently fizzing and giving off hydrogen. It gradually reacts and disappears, forming a colourless solution of lithium hydroxide. The reaction generates heat too slowly and lithium's melting point is too high for it to melt.

Sodium

Sodium also floats on the surface, but enough heat is given off to melt the sodium (sodium has a lower melting point than lithium and the reaction produces heat faster) and it melts almost at once to form a small silvery ball that dashes around the surface. A white trail of sodium hydroxide is seen in the water under the sodium, but this soon dissolves to give a colourless solution of sodium hydroxide.The sodium moves because it is pushed around by the hydrogen which is given off during the reaction. If the sodium becomes trapped on the side of the container, the hydrogen may catch fire to burn with an orange flame. The colour is due to contamination of the normally blue hydrogen flame with sodium compounds.

Potassium

Potassium behaves rather like sodium except that the reaction is faster and enough heat is given off to set light to the hydrogen. This time the normal hydrogen flame is contaminated by potassium compounds and so is coloured lilac (a faintly bluish pink).

Rubidium

Rubidium is denser than water and so sinks. It reacts violently and immediately, with everything spitting out of the container again. Rubidium hydroxide solution and hydrogen are formed.

Caesium

Caesium explodes on contact with water, quite possibly shattering the container. Caesium hydroxide and hydrogen are formed

The Group 1 metals become more reactive towards water as you go down the Group.

The overall enthalpy changes

The table gives estimates of the enthalpy change for each of the elements undergoing the reaction:

$$X_{(s)} + H_2O_{(l)} \longrightarrow XOH_{(aq)} + \frac{1}{2} H_{2(g)}$$

	enthalpy change (kJ / mol)
Li	-222
Na	-184
K	-196
Rb	-195
Cs	-203

When these reactions happen, the differences between them lie entirely in what is happening to the metal atoms present. In each case, you start with metal atoms in a solid and end up with metal ions in solution.Overall, what happens to the metal is this:

$$X_{(s)} \longrightarrow X^{+}_{(aq)} + e^{-}$$

You can calculate the overall enthalpy change for this process by using Hess's Law and breaking it up into several steps that we know the enthalpy changes for.First, you would need to supply *atomisation energy* to give gaseous atoms of the metal.

$$X_{(s)} \longrightarrow X_{(g)}$$

Then ionise the metal by supplying its first *ionisation energy.*

$$X_{(g)} \longrightarrow X^{+}_{(g)} + e^{-}$$

And finally, you would get *hydration enthalpy* released when the gaseous ion comes into contact with water.

$$X^{+}_{(g)} + (aq) \longrightarrow X^{+}_{(aq)}$$

If we put values for all these steps into a table, they look like this (all values in kJ / mol):

	at. Energy	1st IE	hydr. Enthalpy	total
Li	+161	+519	-519	+161
Na	+109	+494	-406	+197
K	+90	+418	-322	+186
Rb	+86	+402	-301	+187
Cs	+79	+376	-276	+179

The atomisation energy is a measure of the strength of the metallic bond in each element. This is falling as the atom gets bigger and the metallic bond is getting longer. The delocalized electrons are further from the attraction of the nuclei in the bigger atoms.The first ionisation energy is falling because the electron being removed is getting more distant from the.nucleus.The extra protons in the nucleus are screened by additional layers of electrons.The hydration enthalpy is a measure of the attraction between the metal ions and lone pairs on water molecules. As the ions get bigger, the water molecules are further from the attraction of the nucleus. The extra protons in the nucleus are again screened by the extra layers of electrons.What is happening is that the various factors are falling at different rates. That destroys any overall pattern.It is, however, possible to look at the table again and find a pattern which is useful.Let's take the last table and just look at the energy input terms - the two processes where you have to supply energy to make them work. In other words, we will miss out the hydration enthalpy term and just add up the other two.

	at. energy	1st IE	total
Li	+161	+519	+680
Na	+109	+494	+603
K	+90	+418	+508
Rb	+86	+402	+488
Cs	+79	+376	+455

There is a steady fall as you go down the Group. As you go from lithium to caesium, you need to put less energy into the reaction to get a positive ion formed. This energy will be recovered later on, but has to be supplied initially. This is going to be related to the activation energy of the reaction.The lower the activation energy, the faster the reaction.So although lithium releases most heat during the reaction, it does it relatively slowly - it isn't all released in one short, sharp burst. Caesium, on the other hand, has a significantly lower activation energy, and so although it doesn't release quite as much heat overall, it does it extremely quickly - and you get an explosion.The reactions become easier as the energy needed to form positive ions falls. This is in part due to a decrease in ionisation energy as you go down the Group, and in part to a fall in atomisation energy reflecting weaker metallic bonds as you go from lithium to caesium. This leads to lower activation energies, and therefore faster reactions.

REACTIONS WITH AIR OR OXYGEN

These are all very reactive metals and have to be stored out of contact with air to prevent their oxidation. Reactivity increases as you go down the Group.Lithium, sodium and potassium are stored in oil. (Lithium in fact floats on the oil, but there will be enough oil coating it to give it some protection. It is, anyway, less reactive than the rest of the Group.)Rubidium and caesium are normally stored in sealed glass tubes to prevent air getting at them. They are stored either in a vacuum or in an inert atmosphere of, say, argon. The tubes are broken open when the metal is used.Depending on how far down the Group you are, different kinds of oxide are formed when the metals burn. Reaction with oxygen is just a more dramatic version of the reaction with air.Lithium is unique in the Group because it also reacts with the nitrogen in the air to form lithium nitride.

Lithium

Lithium burns with a strongly red-tinged flame if heated in air. It reacts with oxygen in the air to give white lithium oxide. With pure oxygen, the flame would simply be more intense.

$$4Li + O_2 \longrightarrow 2Li_2O$$

For the record, it also reacts with the nitrogen in the air to give lithium nitride. Lithium is the only element in this Group to form a nitride in this way.

$$6Li + N_2 \longrightarrow 2Li_3N$$

Sodium

Small pieces of sodium burn in air with often little more than an orange glow. Using larger amounts of sodium or burning it in oxygen gives a strong orange flame. You get a white solid mixture of sodium oxide and sodium peroxide.The equation for the formation of the simple oxide is just like the lithium one.

$$4Na + O_2 \longrightarrow 2Na_2O$$

The peroxide equation is:

$$2Na + O_2 \longrightarrow Na_2O_2$$

Potassium

Small pieces of potassium heated in air tend to just melt and turn instantly into a mixture of potassium peroxide and potassium superoxide without any flame being seen. Larger pieces of potassium burn with a lilac flame.The equation for the formation of the peroxide is just like the sodium one above:

$$2K + O_2 \longrightarrow K_2O_2$$

. . . and for the superoxide:

$$K + O_2 \longrightarrow KO_2$$

Rubidium and caesium

Both metals catch fire in air and produce superoxides, RbO_2 and CsO_2. The equations are the same as the equivalent potassium one.Both superoxides are described in most sources as being either orange or yellow.

Lithium (and to some extent sodium) form simple oxides, X_2O, which contain the common O^{2-} ion.

Sodium (and to some extent potassium) form peroxides, X_2O_2, containing the more complicated O_2^{2-} ion.

Potàssium, rubidium and caesium form superoxides, XO_2. The structure of the superoxide ion, O_2^-, is too difficult to discuss at this level, needing a good knowledge of molecular orbital theory to make sense of it.

The more complicated ions aren't stable in the presence of a small positive ion. Consider the peroxide ion, for example.

The peroxide ion, O_2^{2-} looks like this:

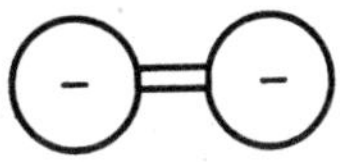

Two oxygen atoms each carrying a negative change, and joind by a covalent bond.

The covalent bond between the two oxygen atoms is relatively weak.

Now imagine bringing a small positive ion close to the peroxide ion. Electrons in the peroxide ion will be strongly attracted towards the positive ion. This is then well on the way to forming a simple oxidc ion if thc right-hand oxygen atom breaks off.

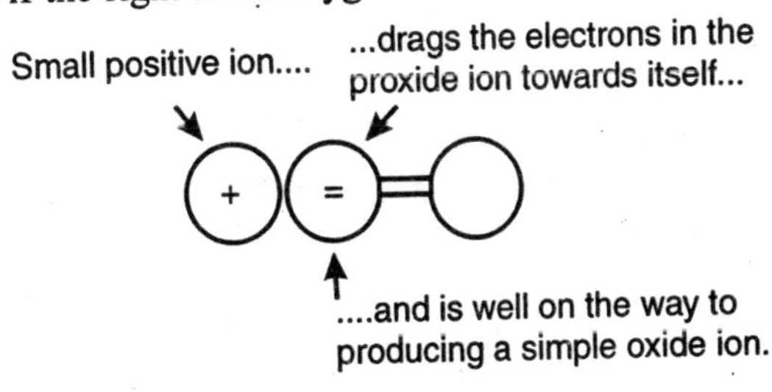

We say that the positive ion *polarises* the negative ion. This works best if the positive ion is small and highly charged - if it has a *high charge density*.Even though it only has one charge, the lithium ion at the top of the Group is so small and has such a high charge density that any peroxide ion near it falls to pieces to give an oxide and oxygen. As you go down the Group to sodium and potassium the positive ions get bigger and they don't have so much effect on the peroxide ion.The superoxide ions are even more easily pulled apart, and these are only stable in the presence of the big ions towards the bottom of the Group.So why do any of the metals form the more complicated oxides? It is a matter of energetics.In the presence of sufficient oxygen, they produce the compound whose formation gives out most energy. That gives the most stable compound.The amount of heat evolved per mole of rubidium in forming its various oxides is:

enthalpy change (kJ / mol of Rb)

Rb_2O -169.5,

Rb_2O_2-236

RbO_2 -278.7

The values for the various potassium oxides show exactly the same trends. As long as you have enough oxygen, forming the peroxide releases more energy per mole of metal than forming the simple oxide. Forming the superoxide releases even more.Forming the more complicated oxides from the metals releases more energy and makes the system more energetically stable. BUT . . . this only works for the metals in the lower half of the Group where the metal ions are big and have a low charge density.At the top of the Group, the small ions with a higher charge density tend to polarise the more complicated oxide ions to the point of destruction.

Reaction with Water

These are simple basic oxides, reacting with water to give the metal hydroxide.For example, lithium oxide reacts with water to give a colourless solution of lithium hydroxide.

$$X_2O + H_2O \longrightarrow 2XOH$$

Reaction with Dilute Acids

These simple oxides all react with an acid to give a salt and water. For example, sodium oxide will react with dilute hydrochloric acid to give colourless sodium chloride solution and water.

$$X_2O + 2HCl \longrightarrow 2XCl + H_2O$$

Reaction with Water

If the reaction is done ice cold (and the temperature controlled so that it doesn't rise even though these reactions are strongly exothermic), a solution of the metal hydroxide and hydrogen peroxide is formed.

$$X_2O_2 + 2H_2O \longrightarrow 2XOH + H_2O_2$$

If the temperature increases (as it inevitably will unless the peroxide is added to water very, very, very slowly!), the hydrogen peroxide produced decomposes into water and oxygen. The reaction can be very violent overall.

Reaction with Dilute Acids

These reactions are even more exothermic than the ones with water. A solution containing a salt and hydrogen peroxide is formed. The hydrogen peroxide will decompose to give water and oxygen if the temperature rises - again, it is almost impossible to avoid this. Another potentially violent reaction!

$$X_2O_2 + 2HCl \longrightarrow 2XCl + H_2O_2$$

Reaction with Water

This time, a solution of the metal hydroxide and hydrogen peroxide is formed, but oxygen gas is given off as well. Once again, these are strongly exothermic reactions and the heat produced will inevitably decompose the hydrogen peroxide to water and more oxygen. Again violent!

$$2XO_2 + 2H_2O \longrightarrow 2XOH + H_2O_2 + O_2$$

Reaction with Dilute Acids

Again, these reactions are even more exothermic than the ones with water. A solution containing a salt and hydrogen peroxide is formed together with oxygen gas. The hydrogen peroxide will again decompose to give water and oxygen as the temperature rises. Violent!

$$2XO_2 + 2HCl \longrightarrow 2XCl + H_2O_2 + O_2$$

The Reactions of the Elements with Chlorine

This is included because of the similarity in appearance between the reactions of the Group 1 metals with chlorine and with oxygen.Sodium, for example, burns with an intense orange flame in chlorine in exactly the same way that it does in pure oxygen. The rest also behave the same in both gases.In each case, there is a white solid residue which is the simple chloride, XCl. There is nothing in any way complicated about these reactions!

$$2X + Cl_2 \longrightarrow 2XCl$$

Some compounds of the Group 1 elements (lithium, sodium, potassium, rubidium and caesium). You will find some information about the nitrates, carbonates, hydrogencarbonates and hydrides of the metals.We will first look at what happens to some of the compounds on heating, and then their solubility.

Group 1 compounds are more stable to heat than the corresponding compounds in Group 2. That the lithium compounds behave similarly to Group 2 compounds, but the rest of Group 1 are in some way different.

Heating the Nitrates

Most nitrates tend to decompose on heating to give the metal oxide, brown fumes of nitrogen dioxide, and oxygen.

For example, a typical *Group 2 nitrate* like magnesium nitrate decomposes like this:

$$2Mg(NO_3)_{2(s)} \longrightarrow 2MgO_{(s)} + 4NO_{2(g)} + O_{2(g)}$$

In *Group 1*, lithium nitrate behaves in the same way - producing lithium oxide, nitrogen dioxide and oxygen.

$$4LiNO_{3(s)} \longrightarrow 2Li_2O_{(s)} + 4NO_{2(g)} + O_{2(g)}$$

The rest of the Group, however, don't decompose so completely (at least not at Bunsen temperatures) - producing the metal nitrite and oxygen, but no nitrogen dioxide.

$$2XNO_{3(s)} \longrightarrow 2XNO_{2(s)} + O_{2(g)}$$

All the nitrates from sodium to caesium decomposes in this same way, the only difference being how hot they have to be to undergo the reaction. As you go down the Group, the decomposition gets more difficult, and you have to use higher temperatures.

Heating the Carbonates

Most carbonates tend to decompose on heating to give the metal oxide and carbon dioxde.For example, a typical *Group 2 carbonate* like calcium carbonate decomposes like this:

$$CaCO_{3(s)} \longrightarrow CaO_{(s)} + CO_{2(g)}$$

In *Group 1*, lithium carbonate behaves in the same way - producing lithium oxide and carbon dioxide.

$$Li_2CO_{3(s)} \longrightarrow Li_2O_{(s)} + CO_{2(g)}$$

The rest of the Group 1 carbonates don't decompose at Bunsen temperatures, although at higher temperatures they will. The decomposition temperatures again increase as you go down the Group.

The Thermal Stability of the Hydrogencarbonates

The *Group 2* hydrogencarbonates like calcium hydrogencarbonate are so unstable to heat that they only exist in solution. Any attempt to get them out of solution causes them to decompose to give the carbonate, carbon dioxide and water.

$$Ca(HCO_3)_{2(aq)} \longrightarrow CaCO_{3(s)} + CO_{2(g)} + H_2O_{(l)}.$$

By contrast, the *Group 1* hydrogencarbonates are stable enough to exist as solids, although they do decompose easily on heating. For example, for sodium hydrogencarbonate:

$$2NaHCO_{3(s)} \longrightarrow Na_2CO_{3(s)} + CO_{2(g)} + H_2O_{(l)}$$

Detailed explanations are given for the carbonates because the diagrams are easier to draw. Exactly the same arguments apply to the nitrates or hydrogencarbonates.There are two ways of explaining the increase in thermal stability as you go down the Group. The hard way is in terms of the energetics of the process; the simple way is to look at the polarising ability of the positive ions.A small positive ion has a lot of charge packed into a small volume of space - especially if it has more than one positive charge. It has a high *charge density* and will have a marked distorting effect on any negative ions which happen to be near it.A bigger positive ion has the same charge spread over a larger volume of space. Its charge density will be lower, and it will cause less distortion to nearby negative ions.

The Carbonates

For example, *Group 2* carbonates are virtually insoluble in water. Magnesium carbonate is soluble to the extent of about 0.02 g per 100 g of water at room temperature.By contrast, the *least* soluble *Group 1* carbonate is lithium carbonate. A saturated solution of it has a concentration of about 1.3 g per 100 g of water at 20°C. The other carbonates in the Group all count as very soluble - increasing to an astonishing 261.5 g per 100 g of water at this temperature for caesium carbonate.Solubility of the carbonates increases as you go down Group 1.

THE HYDROXIDES

The least soluble hydroxide in *Group 1* is lithium hydroxide - but it is still possible to make a solution with a concentration of 12.8 g per 100 g of water at 20°C. The other hydroxides in the Group are even more soluble.Solubility of the hydroxides increases as you go down Group 1.In *Group 2*, the *most* soluble one is barium hydroxide - and it is only possible to make a solution of concentration around 3.9 g per 100 g of water at the same temperature.Explaining the trends in Group 2 was difficult enough. Comparing them with Group 1 is going to be even more difficult - particularly in the case of the carbonates, because the trends in the two Groups are in opposite directions. The carbonates get more soluble as you go down Group 1, but tend to get less soluble down Group 2.

Saline (salt-like) Hydrides

The hydrides of Group 1 metals are white crystalline solids which contain the metal ions and hydride ions, H^-. They have exactly the same crystal structure as sodium chloride - that's why they are called saline or salt-like hydrides.Because they can react violently with water or moist air, they are normally supplied as suspensions in mineral oil.

Preparation of the Group 1 hydrides

These are made by passing hydrogen gas over the heated metal. For example, for lithium hydride:

$$2Li + H_2 \longrightarrow 2LiH$$

Reactions of the Group 1 hydrides

ELECTROLYSIS

On heating, most of these hydrides decompose back into the metal and hydrogen before they melt. It is, however, possible to melt lithium hydride and to electrolyse the melt.The metal is released at the cathode as you would expect. Hydrogen is given off at the anode (the positive electrode) and this is evidence for the presence of the negative hydride ion in lithium hydride.The anode equation is:

$$2H^- \longrightarrow H_2 + 2e^-$$

The other Group 1 hydrides can be electrolysed in solution in various molten mixtures such as a mixture of lithium chloride and potassium chloride. Mixtures such as these melt at lower temperatures than the pure chlorides.

Reaction with Water

These hydrides react violently with water releasing hydrogen gas and producing the metal hydroxide.For example, sodium hydride reacts with water to produce a solution of sodium hydroxide and hydrogen gas.

$$NaH + H_2O \longrightarrow NaOH + H_2$$

FLAME TESTS

A flame test for a range of metal ions, and briefly describes how the flame colour arises.Flame tests are used to identify the presence of a relatively small number of metal ions in a compound. Not all metal ions give flame colours.For Group 1 compounds, flame tests are usually by far the easiest way of identifying which metal you have got. For other metals, there are usually other easy methods which are more reliable - but the flame test can give a useful hint as to where to look.

Carrying out a Flame Test

Clean a platinum or nichrome (a nickel-chromium alloy) wire by dipping it into concentrated hydrochloric acid and then holding it in a hot Bunsen flame. Repeat this until the wire doesn't produce any colour in the flame.When the wire is clean, moisten it again with some of the acid and then dip it into a small amount of the solid you are testing so that some sticks to the wire. Place the wire back in the flame again.If the flame colour is weak, it is often worthwhile to dip the wire back in the acid again and put it back into the flame as if you were cleaning it.

The Colours

The colours in the table are just a guide. Almost everybody sees and describes colours differently. for example, used the word "red" several times to describe colours which can be quite different from each other.

	flame colour
Li	red
Na	strong persistent orange
K	lilac (pink)
Rb	red (reddish-violet)
Cs	blue violet
Ca	orange-red
Sr	red
Ba	pale green
Cu	blue-green (often with white flashes)
Pb	greyish-white

The Origin of Flame Colours

Flame colours are produced from the movement of the electrons in the metal ions present in the compounds.

For example, a sodium ion in an unexcited state has the structure $1s^22s^22p^6$.When you heat it, the electrons gain energy and can jump into any of the empty orbitals at higher levels - for example, into the 7s or 6p or 4d or whatever, depending on how much energy a particular electron happens to absorb from the flame.Because the electrons are now at a higher and more energetically unstable level, they tend to fall back down to where they were before - but not necessarily all in one go.

An electron which had been excited from the 2p level to an orbital in the 7 level, for example, might jump back to the 2p level in one go. That would release a certain amount of energy which would be seen as light of a particular colour.

However, it might jump back in two (or more) stages. For example, first to the 5 level and then back to the 2 level.

Each of these jumps involves a specific amount of energy being released as light energy, and each corresponds to a particular colour.As a result of all these jumps, a spectrum of coloured lines will be produced. The colour you see will be a combination of all these individual colours.The exact sizes of the possible jumps in energy terms vary from one metal ion to another. That means that each different ion will have a different pattern of spectral lines, and so a different flame colour.

GROUP 2 (ALKALINE EARTH METALS)

The trends in some atomic and physical properties of the Group 2 elements - beryllium, magnesium, calcium, strontium and barium. The trends in atomic radius, first ionisation energy, electronegativity and melting point.Even if you aren't currently interested in all these things.The same ideas tend to recur throughout the atomic properties.

Atomic Radii of the Group 2 elements

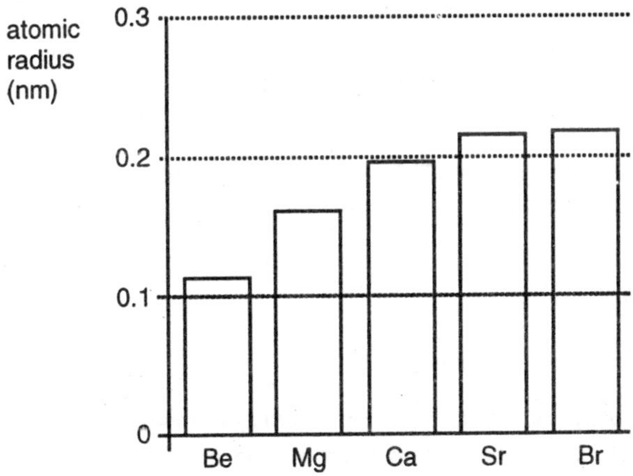

The atomic radius increases as you go down the Group. Notice that beryllium has a particularly small atom compared with the rest of the Group.The radius of an atom is governed by

1. The number of layers of electrons around the nucleus
2. The pull the outer electrons feel from the nucleus.

Compare beryllium and magnesium:

Be $1s^2 2s^2$

Mg $1s^2 2s^2 2p^6 3s^2$

In each case, the two outer electrons feel a net pull of 2+ from the nucleus. The positive charge on the nucleus is cut down by the negativeness of the inner electrons.

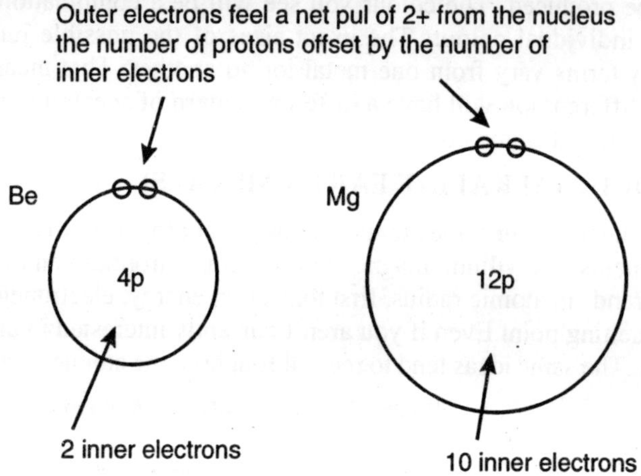

This is equally true for all the other atoms in Group 2. Work it out for calcium if you aren't convinced.

The only factor which is going to affect the size of the atom is therefore the number of layers of inner electrons which have to be fitted in around the atom. Obviously, the more layers of electrons you have, the more space they will take up - electrons repel each other. That means that the atoms are bound to get bigger as you go down the Group.

First Ionisation Energy

First ionisation energy is the energy needed to remove the most loosely held electron from each of one mole of gaseous atoms to make one mole of singly charged gaseous ions - in other words, for 1 mole of this process:

$$X_{(g)} \longrightarrow X^{+}_{(g)} + e^{-}$$

First Ionisation Energy of the Group 2 elements

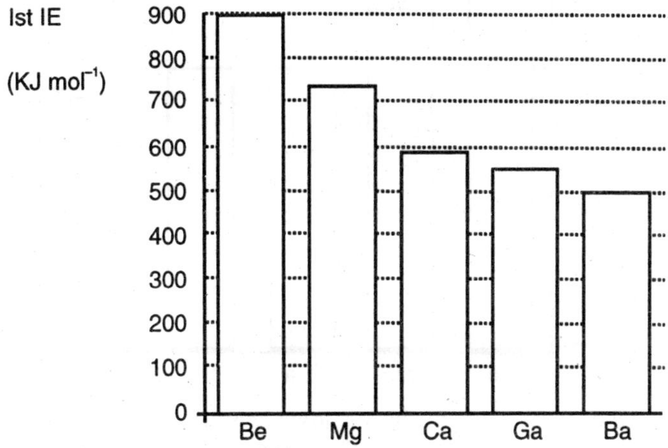

Notice that first ionisation energy falls as you go down the group.Ionisation energy is governed by

1. The charge on the nucleus,
2. The amount of screening by the inner electrons,
3. The distance between the outer electrons and the nucleus.

As you go down the Group, the increase in nuclear charge is exactly offset by the increase in the number of inner electrons. Just as when we were talking about atomic radius further up, in each of the elements in this Group, the outer electrons feel a net attraction of 2+ from the centre.However, as you go down the Group, the distance between the nucleus and the outer electrons increases and so they become easier to remove - the ionisation energy falls.

ELECTRONEGATIVITY

Electronegativity is a measure of the tendency of an atom to attract a bonding pair of electrons. It is usually measured on the

Pauling scale, on which the most electronegative element (fluorine) is given an electronegativity of 4.0.

Electronegativity of the Group 2 elements

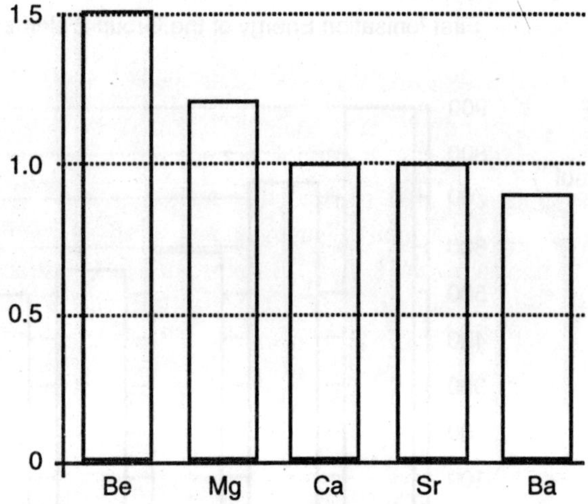

All of these elements have a low electronegativity. (Remember that the most electronegative element, fluorine, has an electronegativity of 4.0.) Notice that electronegativity falls as you go down the Group. The atoms become less and less good at attracting bonding pairs of electrons.Imagine a bond between a magnesium atom and a chlorine atom. Think of it to start with as a covalent bond - a pair of shared electrons. The electron pair will be dragged towards the chlorine end because there is a much greater net pull from the chlorine nucleus than from the magnesium one.

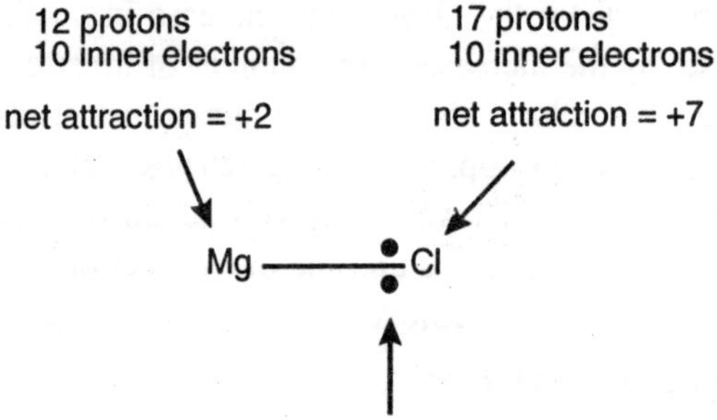

The electron pair ends up so close to the chlorine that there is essentially a transfer of an electron to the chlorine - ions are formed.The large pull from the chlorine nucleus is why chlorine is much more electronegative than magnesium is.Now compare this with the beryllium-chlorine bond.The net pull from each end of the bond is the same as before, but you have to remember that the beryllium atom is smaller than a magnesium atom. That means that the electron pair is going to be closer to the net 2+ charge from the beryllium end, and so more strongly attracted to it.

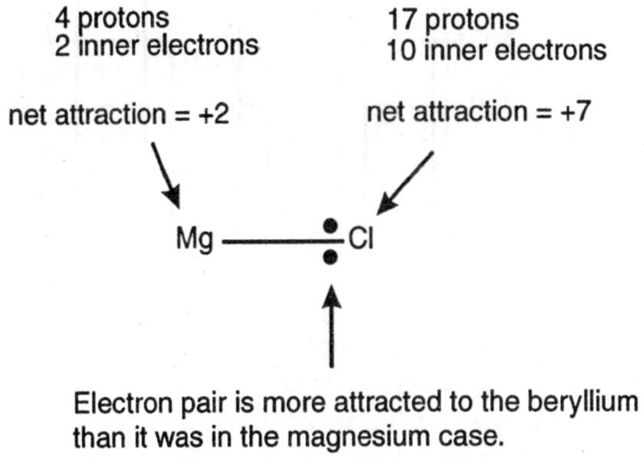

In this case, the electron pair doesn't get attracted close enough to the chlorine for an ionic bond to be formed. Because of its small size, beryllium forms covalent bonds, not ionic ones. The attraction between the beryllium nucleus and a bonding pair is always too great for ions to be formed.

As the metal atoms get bigger, any bonding pair gets further and further away from the metal nucleus, and so is less strongly attracted towards it. In other words, as you go down the Group, the elements become less electronegative.As you go down the Group, the bonds formed between these elements and other things such as chlorine become more and more ionic. The bonding pair is increasingly attracted away from the Group 2 element towards the chlorine.

Melting Point

With the exception of magnesium the melting point falls as you go down the Group. It would be tempting to say that apart from magnesium the lower melting points as you go down the Group reflect weaker metallic bonds, but this is dangerous.

Melting Points of the Group 2 elements

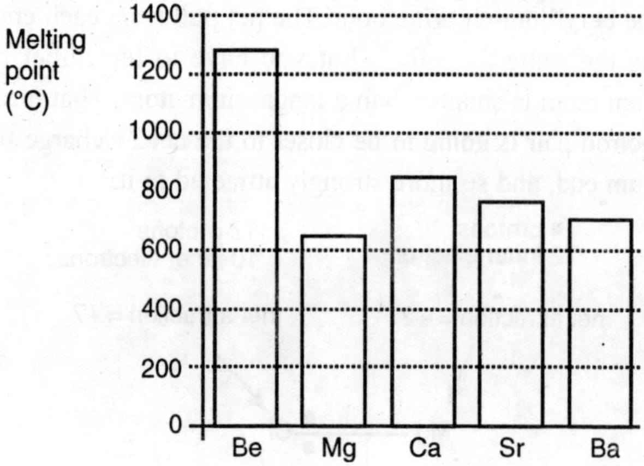

The metallic bond isn't fully broken on melting. Boiling point is usually a better guide to the strengths of the forces involved.You will sometimes come across an explanation for the low melting point of magnesium in terms of the way the atoms pack in the crystal. It isn't very difficult to pick holes in this argument and, in any case, the boiling point of magnesium is also surprisingly low - and packing is completely irrelevant when you are thinking about a liquid. In the case of magnesium, there must be some factor producing a relatively weak metallic bond.

Boilling Points of the Group 2 elements

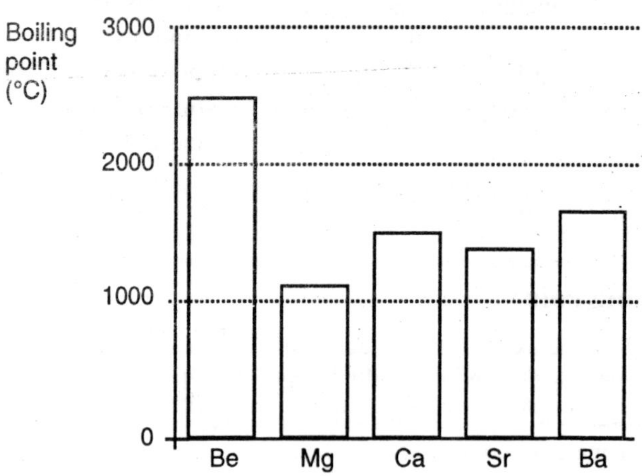

You will see that there is no obvious pattern in the boiling points - implying that there is no simple pattern in the strengths of the metallic bonds.Another measure of the strengths of the metallic bonds might be atomisation energy. This is the energy needed to produce 1 mole of separated atoms in the gas state starting from the element in its standard state.

Atomisation Energy of the Group 2 elements

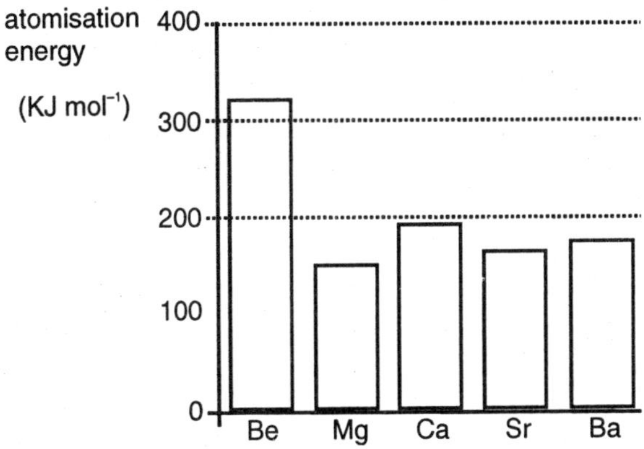

Reactions with Water

Beryllium, magnesium, calcium, strontium and barium - with water (or steam). It uses these reactions to explore the trend in reactivity in Group 2.

Beryllium

Beryllium has no reaction with water or steam even at red heat.

Magnesium

Magnesium burns in steam to produce magnesium oxide and hydrogen.

$$Mg_{(s)} + H_2O_{(g)} \longrightarrow MgO_{(s)} + H_{2(g)}$$

Very clean magnesium has a very slight reaction with cold water. The reaction soon stops because the magnesium hydroxide formed is almost insoluble in water and forms a barrier on the magnesium preventing further reaction.

$$Mg_{(s)} + 2H_2O_{(l)} \longrightarrow Mg(OH)_{2(s)} + H_{2(g)}$$

Calcium, strontium and barium

These all react with cold water with increasing vigour to give the metal hydroxide and hydrogen. Strontium and barium have reactivities similar to lithium in Group 1 of the Periodic Table.The equation for the reactions of any of these metals would be:

$$X_{(s)} + 2H_2O_{(l)} \longrightarrow X(OH)_{2(aq\ or\ s)} + H_{2(g)}$$

The hydroxides aren't very soluble, but they get more soluble as you go down the Group. The calcium hydroxide formed shows up mainly as a white precipitate (although some does dissolve). You get less precipitate as you go down the Group because more of the hydroxide dissolves in the water.

The Group 2 metals become more reactive towards water as you go down the Group.

Beryllium There is an additional reason for the lack of reactivity of beryllium compared with the rest of the Group. Beryllium has a strong resistant layer of oxide on its surface which lowers its reactivity. If you add that to the trends explained below, beryllium turns out to be very unreactive.

The enthalpy change of a reaction is a measure of the amount of heat absorbed or evolved when the reaction takes place. An enthalpy change is negative if heat is evolved, and positive if it is absorbed. If you calculate the enthalpy change for the possible reactions between beryllium or magnesium and steam, you come up with these answers:

$$Be_{(s)} + H_2O_{(g)} \longrightarrow BeO_{(s)} + H_{2(g)} \qquad \Delta H = -369 \text{ kJ mol}^{-1}$$

$$Mg_{(s)} + H_2O_{(g)} \longrightarrow MgO_{(s)} + H_{2(g)} \qquad \Delta H = -360 \text{ kJ mol}^{-1}$$

Notice that both possible reactions are strongly exothermic, giving out almost identical amounts of heat. However, only the magnesium reaction actually happens. The explanation for the different reactivities must lie somewhere else.Similarly, if you calculate the enthalpy changes for the reactions between calcium, strontium or barium and cold water, you again find that the amount of heat evolved in each case is almost exactly the same - in this case, about -430 kJ mol^{-1}.The reason for the increase in reactivity must again lie elsewhere.The activation energy for a reaction is the minimum amount of energy which is needed in order for the reaction to take place. It doesn't matter how exothermic the reaction would be once it got started - if there is a high activation energy barrier, the reaction will take place very slowly, if at all.When Group 2 metals react to form oxides or hydroxides, metal ions are formed.

The formation of the ions from the original metal involves various stages all of which require the input of energy - contributing to the activation energy of the reaction. These stages involve the input of:

1. The atomisation energy of the metal. This is the energy needed to break the bonds holding the atoms together in the metallic lattice.
2. The first + second ionisation energies. These are necessary to convert the metal atoms into ions with a 2+ charge.

After this, there will be a number of steps which give out heat again - leading to the formation of the products, and overall exothermic reactions.The graph shows the effect of these important energy-absorbing stages as you go down Group 2.

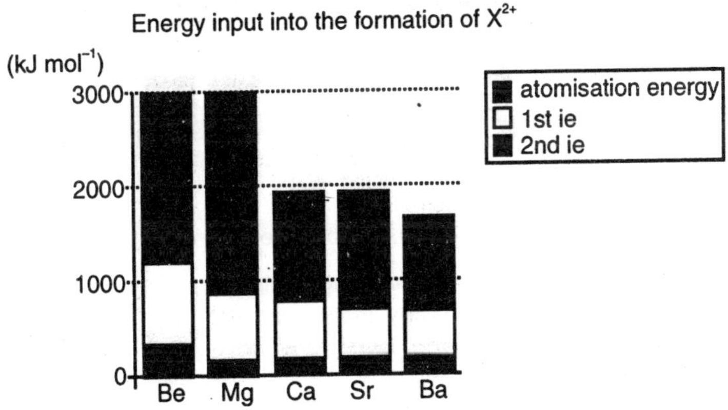

Notice that the ionisation energies dominate this - particularly the second ionisation energies. Ionisation energies fall as you go down the Group. Because it gets easier to form the ions, the reactions will happen more quickly.The reactions become easier as the energy needed to form positive ions falls. This is mainly due to a decrease in ionisation energy as you go down the Group. This leads to lower activation energies, and therefore faster reactions.

THE REACTIONS WITH OXYGEN

Formation of Simple Oxides

On the whole, the metals burn in oxygen to form a simple metal oxide.Beryllium is reluctant to burn unless it is in the form of dust or powder. Beryllium has a very strong (but very thin) layer of beryllium oxide on its surface, and this prevents any new oxygen getting at the underlying beryllium to react with it.

$$2X_{(s)} + O_{2(g)} \longrightarrow 2XO_{(s)}$$

"X" in the equation can represent any of the metals in the Group.It is almost impossible to find any trend in the way the metals react with oxygen. It would be quite untrue to say that they burn more vigorously as you go down the Group.To be able to make any sensible comparison, you would have to have pieces of metal which were all equally free of oxide coating, with exactly the same surface area and shape, exactly the same flow of oxygen around them, and heated to exactly the same extent to get them started.

Beryllium: beryllium burns with the same sort of silvery sparkles that magnesium or aluminium powder burn with if they are scattered into a flame.

Magnesium: of course, burns with a typical intense white flame.

Calcium: is quite reluctant to start burning, but then bursts dramatically into flame, burning with an intense white flame with a tinge of red at the end.

Strontium: burns with a much richer red tinge to the flame.

Barium: the flame appeared to be white with some pale green tinges. It wasn't noticeably any more dramatic than the familiar magnesium flame.

Formation of Peroxides

Strontium and barium will also react with oxygen to form strontium or barium peroxide.Strontium forms this if it is heated in oxygen under high pressures, but barium forms barium peroxide just on normal heating in oxygen. Mixtures of barium oxide and barium peroxide will be produced.

$$Ba_{(s)} + O_{2(g)} \longrightarrow BaO_{2(s)}$$

The strontium equation would look just the same.

The Reactions with Air

The reactions of the Group 2 metals with air rather than oxygen is complicated by the fact that they all react with nitrogen to produce nitrides. In each case, you will get a mixture of the metal oxide and the metal nitride.

The general equation for the Group is:

$$3X_{(s)} + N_{2(g)} \longrightarrow X_3N_{2(s)}$$

The familiar white ash you get when you burn magnesium ribbon in air is a mixture of magnesium oxide and magnesium nitride There are no simple patterns. It would be tempting to say that the reactions get more vigorous as you go down the Group, but it isn't true.The overall amount of heat evolved when one mole of oxide is produced from the metal and oxygen shows no simple pattern:

Heat evolved when 1 mole of oxide is formed form its elements

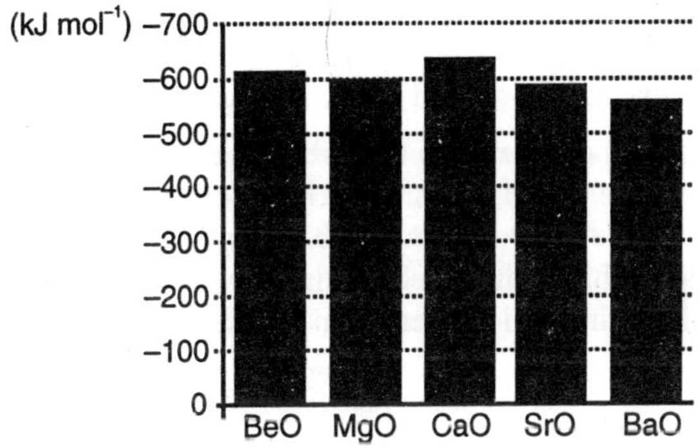

If anything, there is a slight tendency for the amount of heat evolved to get less as you go down the Group.But how reactive a metal seems to be depends on how fast the reaction happens - not the overall amount of heat evolved. The speed is controlled by factors like the presence of surface coatings on the metal and the size of the activation energy.You could argue that the activation energy will fall as you go down the Group and that will make the reaction go faster. The activation energy will fall because the ionisation energies of the metals fall.In this case, though, the effect of the fall in the activation energy is masked by other factors - for example, the presence of existing oxide layers on the metals, and the impossibility of controlling precisely how much heat you are supplying to the metal in order to get it to start burning.It is interesting to look at what happens if you heat a *very* reactive metal like potassium in air. The potassium melts at a low temperature and almost instantly turns into a pool of molten potassium oxide. The activation energy is so low that the reaction happens very quickly at quite a low temperature.Magnesium, on the other hand, has to be heated to quite a high temperature before it will start to react. The activation energy is much higher. There are also problems with surface coatings. It is then so hot that it produces the typical intense white flame.It would obviously be totally misleading to say that magnesium is more reactive than potassium on the evidence of the bright flame.

Beryllium, magnesium and calcium don't form peroxides when heated in oxygen, but strontium and barium do. There is an increase in the tendency to form the peroxide as you go down the Group.The peroxide ion, O_2^{2-} looks llike this:

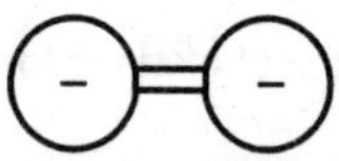

Two oxygen atoms each carrying
a negative change, and joind by a covalent bond.

The covalent bond between the two oxygen atoms is relatively weak.Now imagine bringing a small 2+ ion close to the peroxide ion. Electrons in the peroxide ion will be strongly attracted towards the positive ion. This is then well on the way to forming a simple oxide ion if the right-hand oxygen atom breaks off.

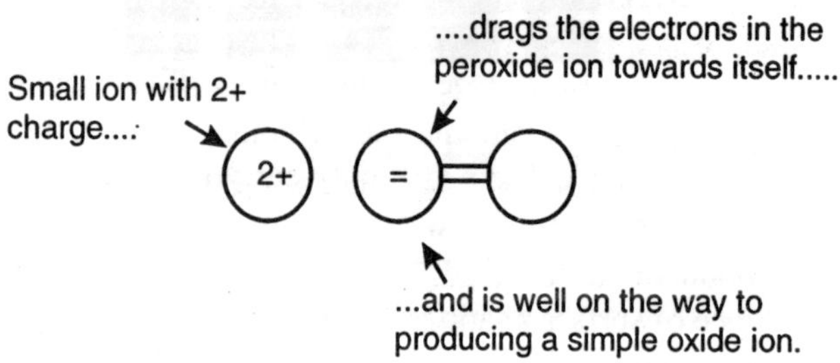

We say that the positive ion *polarises* the negative ion. This works best if the positive ion is small and highly charged - if it has a *high charge density*.Ions of the metals at the top of the Group have such a high charge density (because they are so small) that any peroxide ion near them falls to pieces to give an oxide and oxygen. As you go down the Group and the positive ions get bigger, they don't have so much effect on the peroxide ion.Barium peroxide can form because the barium ion is so large that it doesn't have such a devastating effect on the peroxide ions as the metals further up the Group.

Nitrogen is often thought of as being fairly unreactive, and yet all these metals combine with it to produce nitrides, X_3N_2, containing X^{2+} and N^{3-} ions.Nitrogen is fairly unreactive because of the very large amount of energy needed to break the triple bond joining the two atoms in the nitrogen molecule, N_2.When something like magnesium nitride forms, you have to supply all the energy needed to form the magnesium ions as well as breaking the nitrogen-nitrogen bonds and then forming N^{3-} ions. All of these processes absorb energy.This energy has to be recovered from somewhere to give an overall exothermic reaction - if the energy can't be recovered, the overall change will be endothermic and won't happen.

Energy is evolved when the ions come together to produce the crystal lattice. This energy is known as *lattice energy* or *lattice enthalpy*.The size of the lattice energy depends on the attractions between the ions. The lattice energy is greatest if the ions are small and highly charged - the ions will be close together with very strong attractions. In the whole of Group 2, the attractions between the 2+ metal ions and the 3- nitride ions are big enough to produce very high lattice energies.When the crystal lattices form, so much energy is released that it more than compensates for the energy needed to produce the various ions in the first place. The excess energy evolved makes the overall process exothermic.This is in contrast to what happens in Group 1 of the Periodic Table (lithium, sodium, potassium, rubidium and caesium). Their ions only carry one positive charge, and so the lattice energies of their nitrides will be much less.Lithium is the only metal in Group 1 to form a nitride. Lithium has by far the smallest ion in the Group, and so lithium nitride has the largest lattice energy of any possible Group 1 nitride. Only in lithium's case is enough energy released to compensate for the energy needed to ionise the metal and the nitrogen - and so produce an exothermic reaction overall.In all the other cases in Group 1, the overall reaction would be endothermic. Those reactions don't happen, and the nitrides of sodium and the rest aren't formed.

SOLUBILITY

The solubility in water of the hydroxides, sulphates and carbonates of the Group 2 elements - beryllium, magnesium, calcium, strontium and barium.

The hydroxides become more soluble as you go down the Group.

Magnesium hydroxide appears to be insoluble in water. However, if you shake it with water, filter it and test the pH of the solution, you find that it is slightly alkaline. This shows that there are more hydroxide ions in the solution than there were in the original water. Some magnesium hydroxide must have dissolved.

Calcium hydroxide solution is used as "lime water". 1 litre of pure water will dissolve about 1 gram of calcium hydroxide at room temperature.

Barium hydroxide is soluble enough to be able to produce a solution with a concentration of around 0.1 mol dm^{-3} at room temperature.

Solubility of the Sulphates

The sulphates become less soluble as you go down the Group.The simple trend is true provided you include hydrated beryllium sulphate in it, but not if the beryllium sulphate is anhydrous.Anyhydrous beryllium sulphate, $BeSO_4$, as *insoluble* whereas the hydrated form, $BeSO_4.4H_2O$ is soluble.You are probably familiar with the reaction between magnesium and dilute sulphuric acid to give lots of hydrogen and a colourless solution of *magnesium sulphate*. Notice that you get a solution, not a precipitate. The magnesium sulphate is obviously soluble.You may also remember that *barium sulphate* is formed as a white precipitate during the test for sulphate ions in solution. The ready formation of a precipitate shows that the barium sulphate must be pretty insoluble. In fact, 1 litre of water will only dissolve about 2 mg of barium sulphate at room temperature.

Solubility of the Carbonates

The carbonates tend to become less soluble as you go down the Group.

None of the carbonates is anything more than very sparingly soluble. Magnesium carbonate is soluble. The trend to lower solubility is, however, broken at the bottom of the Group. Barium carbonate is slightly more soluble than strontium sulphate.

Enthalpy Changes

The usual explanation is in terms of the enthalpy changes which occur when an ionic compound dissolves in water. Energy has to be supplied to break up the lattice of ions, and energy is released when these ions form bonds of one sort or another with water molecules.As you go down a Group, the energy needed to break up the lattice falls as the positive ions get bigger. The bigger the ions, the more distance there is between them, and the weaker the forces holding them together.Again as the positive ions get bigger, the energy released as the ions bond to water molecules (their hydration enthalpies) falls as well. Bigger ions aren't so strongly attracted to the water molecules.Since both of these important enthalpy terms fall as you go down the Group, what matters in deciding whether the change becomes more endothermic or more exothermic overall is how fast they fall relative to each other.Here is an example of the sort of calculations you might do to work out the enthalpy change of solution for sodium chloride and potassium chloride.

	latice enthalpy	hydration enthalpy of positive ion	hydration onthalpy of Cl^-	
NaCl	+771	–406	–364	= +1 kJ mol^{-1}
KCl	+701	–322	–364	= + 15 kJ mol^{-1}

tails by 70 tails by 84

The enthalpy of solution changes from NaCl to KCl because the lattice enthalpy and hydration enthalpy of the positive ion fall by different amounts.

1. If the hydration enthalpy falls faster than the lattice enthalpy (as in this case), the net effect is that the overall change becomes more endothermic (or less exothermic in other possible cases where the total enthalpy change turns out to be negative).
2. If the lattice enthalpy falls faster than the hydration enthalpy, the opposite happens - the change will become less endothermic (or more exothermic).

It turns out that the main factor is the size of the negative ion.

1. For small negative ions like hydroxide, the lattice enthalpy falls faster than the hydration enthalpy of the positive ions.

That means that the enthalpy of solution will become less positive (or more negative).

2. For large negative ions like sulphate or carbonate, the hydration enthalpy of the positive ions falls faster than the lattice enthalpy. In this case, the enthalpy of solution will become more positive (or less negative).

With sulphates, for example, the percentage increase in the inter-ionic distance as you go from magnesium to calcium sulphate isn't as great as it would be with a smaller negative ion like hydroxide.Since the percentage increase in inter-ionic distance isn't very great, the change in the lattice enthalpy won't be very great either.

The Relationship Between Enthalpy of Solution and solubility

The assumption is made that the more endothermic (or less exothermic) the enthalpy of solution is, the less soluble the compound.So sulphates and carbonates become less soluble as you go down the Group; hydroxides become more soluble. The enthalpy of solution values for the Group 1 chlorides.

	Calculated value ($kJ\ mol^{-1}$)	**value($kJ\ mol^{-1}$)**
NaCl	+1	+3.9
KCl	+15	+17.2
RbCl	+10	+16.7
CsCl	+5	+17.9

The reasons for the discrepancies lie in the way the numbers are calculated. These are very small numbers worked out from much larger ones. Small uncertainties in those large numbers will cause large swings in the answers.The solubilities of the Group 1 chlorides (in moles of solute saturating 100 g of water at 298 K) compared with their enthalpies of solution are:

	Calculated value ($kJ\ mol^{-1}$)	**Solubility (mol/100 g of water**
NaCl	+3.9	0.615
KCl	+17.2	0.481
RbCl	+16.7	0.781
CsCl	+17.9	1.13

There is no obvious relationship connecting the relative movements of these solubility values with the enthalpy of solution figures. It would be quite untrue to say that the more endothermic the change, the less soluble the compound.The table above illustrates this problem.Taking the sign of enthalpy of solution at face value.

$MgCO_3$ -25.3

$CaCO_3$ -12.3

$SrCO_3$ -3.4

$BaCO_3$ +4.2

Sodium chloride and the other Group 1 chlorides dissolve despite the fact that their enthalpies of solution are positive, and yet magnesium carbonate (and most of the other Group 2 carbonates) are very sparingly soluble, but have exothermic enthalpies of solution.The Group 1 chlorides, you could argue that the figures get progressively less exothermic, and at barium carbonate become endothermic. That would seem to support the decrease in solubility as you go down the Group quite nicely.If you look at the solubility data, the trend is broken at the bottom of the Group. Barium carbonate is more soluble than strontium carbonate.

To get around the problem of many compounds dissolving freely in water despite the fact that their enthalpies of solution are endothermic you have to introduce the concept of *entropy change*.The only way of making sense of entropy without getting bogged down in some serious maths is to think of it as a measure of the amount of disorder in a system. Entropy is given the symbol S. If a system becomes more disordered, then its entropy increases.In order to see whether a change is possible or not, you have to think about a combination of the enthalpy change and the entropy change. These can be combined mathematically to give an important term known as *free energy change*.

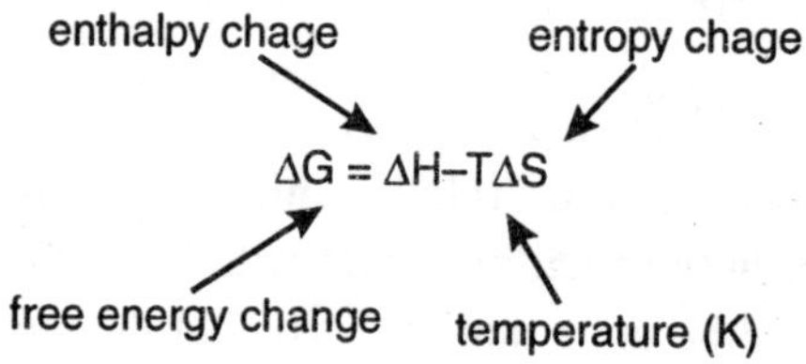

As an approximation, for a reaction to happen, the free energy change must be negative.What happens if the enthalpy change is positive - as for example when sodium chloride dissolves in water.As long as the entropy change is positive enough, it is possible to get a negative value for free energy change. In the sodium chloride case, you don't have to have very much increase in entropy to outweigh the small enthalpy change.

$$\Delta G = \Delta H - T\Delta S$$

If entropy change is positive enough, the negative value of $T\Delta S$ will outweigh the positive value of ΔH.

ΔG will then become negative, and the change is feasible.

So . . . does the entropy increase when sodium chloride dissolve in water? Yes, it does! The sodium and chloride ions were arranged in a very tidy way in the crystal lattice - their entropy was low. When you dissolve the crystal in water, the entropy increases as the ions and water molecules become completely jumbled up - they become much more disordered than they were originally.

The most obvious thing that's wrong is that it won't explain why some compounds (like magnesium carbonate, and most of the other Group 2 carbonates) don't dissolve in water even though their enthalpies of solution are mainly negative. In these cases, the entropy of the system must fall when the compounds dissolve in water - in other words, the solution in water is more ordered than the original crystal and water!

If this is negative, the only way of making ΔG positive is for $-T\Delta G$ to be positive.

$$\Delta G = \Delta H - T\Delta S$$

That only happens if ΔS is negative-if the system becomes more ordered tha it was before.

ΔG must be positive because the change isn't happening to any extent

This happens because the water molecules become more ordered when the compound dissolves in them. Instead of milling around pretty much at random, they become attracted to the ions present and arranged around them. This is particularly effective if the ions are small and highly charged - and so the effect is greatest for the positive ions at the top of the Group, and gets less as you go down. It is also much more important in Group 2 than in Group 1 where the ions only carry one positive charge. That means that you have two entropy effects to consider. There is the increase in disorder as the crystal lattice breaks up, but a corresponding increase in order in the water - which varies depending on the sizes and charges of the ions present. To explain this properly, you need to think about the way lattice enthalpy changes as you go down the Group, the way that hydration enthalpies change, and the way that entropy changes. The way those changes happen will vary from one type of compound to another.

Remember that the solubility of the carbonates falls as you go down Group 2, apart from an increase as you go from strontium to barium carbonate.The general fall is because hydration enthalpies are falling faster than lattice enthalpies. Remember that where you have a big negative ion, its size dominates the inter-ionic distance and so doesn't allow the lattice enthalpy to change much. This gives the enthalpy of solution values we've already looked at (values in kJ mol^{-1}):

$MgCO_3$	-25.3
$CaCO_3$	-12.3
$SrCO_3$	-3.4
$BaCO_3$	+4.2

But the entropy change will also be varying as you go down the Group. At the top, where you have small 2+ ions, the overall entropy change in the system must be negative - the system as a whole becomes more ordered when the compound dissolves because of the way the water molecules become organised around the positive ions. That negative entropy change is going to be enough to wipe out the effect of the exothermic enthalpy of solution.

Towards the bottom of the Group, this effect changes. The bigger ions have less organising effect on the water molecules. The entropy change is becoming less negative (or perhaps even at this stage, positive). That's going to tend to make the compounds more soluble.The overall effect is a complex balance between the way the enthalpy of solution varies and the way the entropy change of solution alters. At barium carbonate, the effect of increasing entropy must be enough to make it more soluble than strontium carbonate.

THERMAL STABILITY

The effect of heat on the carbonates and nitrates of the Group 2 elements - beryllium, magnesium, calcium, strontium and barium. It describes and explains how the thermal stability of the compounds changes as you go down the Group.

All the carbonates in this Group undergo thermal decomposition to give the metal oxide and carbon dioxide. Thermal decomposition is the term given to splitting up a compound by heating it.If "X" represents any one of the elements:

$$XCO_{3(s)} \longrightarrow XO_{(s)} + CO_{2(g)}$$

As you go down the Group, the carbonates have to be heated more strongly before they will decompose.

The carbonates become more stable to heat as you go down the Group.

All the nitrates in this Group undergo thermal decomposition to give the metal oxide, nitrogen dioxide and oxygen.Again, if "X" represents any one of the elements:

$$2X(NO_3)_{2(s)} \longrightarrow 2XO_{(s)} + 4NO_{2(g)} + O_{2(g)}$$

As you go down the Group, the nitrates also have to be heated more strongly before they will decompose.

The nitrates also become more stable to heat as you go down the Group.

Both carbonates and nitrates become more thermally stable as you go down the Group. The ones lower down have to be heated more strongly than those at the top before they will decompose.A small 2+ ion has a lot of charge packed into a small volume of space. It has a high charge density and will have a marked distorting effect on any negative ions which happen to be near it.A bigger 2+ ion has the same charge spread over a larger volume of space. Its charge density will be lower, and it will cause less distortion to nearby negative ions.

Carbonate ion

If you worked out the structure of a carbonate ion using "dots-and-crosses" or some similar method, you would probably come up with:

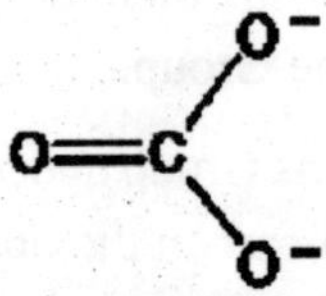

This shows two single carbon-oxygen bonds and one double one, with two of the oxygens each carrying a negative charge. Unfortunately, in real carbonate ions all the bonds are identical, and the charges are spread out over the whole ion - although concentrated on the oxygen atoms. We say that the charges are *delocalised.*This is a rather more complicated version of the bonding you might have come across in benzene or in ions like ethanoate. For the purposes of this topic, you don't need to understand how this bonding has come about.The next diagram shows the delocalised electrons. The shading is intended to show that there is a greater chance of finding them around the oxygen atoms than near the carbon.

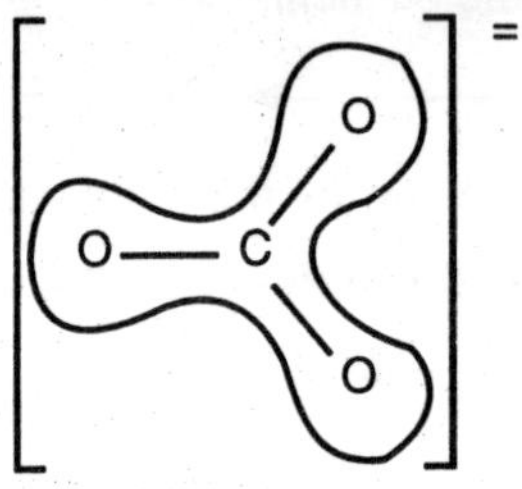

Polarising the Carbonate Ion

Now imagine what happens when this ion is placed next to a positive ion. The positive ion attracts the delocalised electrons in the carbonate ion towards itself. The carbonate ion becomes polarised. The diagram shows what happens with an ion from Group 2, carrying two positive charges.

If this is heated, the carbon dioxide breaks free to leave the metal oxide.How much you need to heat the carbonate before that happens depends on how polarised the ion was. If it is highly polarised, you need less heat than if it is only slightly polarised.If the positive ion only had one positive charge, the polarising effect would be less. That is why the Group 1 compounds are more thermally stable than those in Group 2. You have to heat the Group 1 compound more because the carbonate ions are less polarised by singly charged positive ions. The smaller the positive ion is, the higher the charge density, and the greater effect it will have on the carbonate ion. As the positive ions get bigger as you go down the Group, they have less effect on the carbonate ions near them. To compensate for that, you have to heat the compound more in order to persuade the carbon dioxide to break free and leave the metal oxide.In other words, as you go down the Group, the carbonates become more thermally stable.

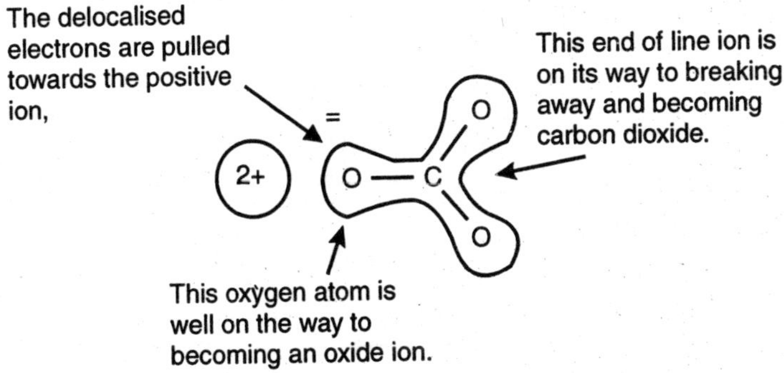

The argument is exactly the same here. The small positive ions at the top of the Group polarise the nitrate ions more than the larger positive ions at the bottom. Drawing diagrams to show this happening is much more difficult because the process has interactions involving more than one nitrate ion. If you calculate the enthalpy changes for the decompostion of the various carbonates, you find that all the changes are quite strongly endothermic. That implies that the reactions are likely to have to be heated constantly to make them happen.The calculated enthalpy changes (in kJ mol^{-1}) are given in the table. Figures to calculate the beryllium carbonate value weren't available. Remember that the reaction we are talking about is:

$$XCO_{3(s)} \longrightarrow XO_{(s)} + CO_{2(g)}$$

$MgCO_3$	+117
$CaCO_3$	+178

$SrCO_3$	+235
$BaCO_3$	+267

The reactions become more endothermic as you go down the Group. That's entirely what you would expect as the carbonates become more thermally stable. You have to supply increasing amounts of heat energy to make them decompose.You can dig around to find the underlying causes of the increasingly endothermic changes as you go down the Group by drawing an enthalpy cycle involving the lattice enthalpies of the metal carbonates and the metal oxides.There are two ways of defining lattice enthalpy. In order to make the argument mathematically simpler.*Lattice enthalpy* is the heat needed to split one mole of crystal in its standard state into its separate gaseous ions. For example, for magnesium oxide, it is the heat needed to carry out 1 mole of this change:

$$MgO_{(s)} \longrightarrow Mg^{2+}_{(g)} + O^{2-}_{(g)} \qquad LE = +3889 \text{ kJ mol}^{-1}$$

The cycle we are interested in looks like this:

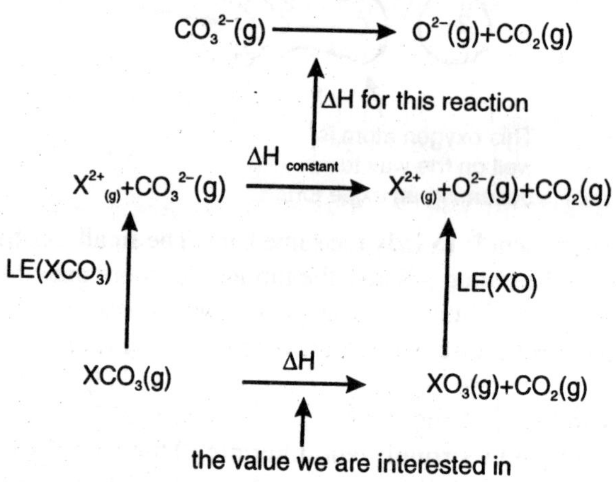

You can apply Hess's Law to this, and find two routes which will have an equal enthalpy change because they start and end in the same places.

For reasons we will look at shortly, the lattice enthalpies of both the oxides and carbonates fall as you go down the Group. But they don't fall at the same rate.The oxide lattice enthalpy falls faster than the carbonate one. If you think carefully about what happens to the value of the overall enthalpy change of the decomposition reaction, you will see that it gradually becomes more positive as you go down the Group.

$$\Delta H = LE(XCO_3) - LE(XO) + \Delta H_{constant}$$

This is a positive number.

The number you are taking away is failing faster than the first number. That means that the overall answer will get more positive.

lattice enthalpy

The size of the lattice enthalpy is governed by several factors, one of which is the distance between the centres of the positive and negative ions in the lattice. Forces of attraction are greatest if the distances between the ions are small. If the attractions are large, then a lot of energy will have to be used to separate the ions - the lattice enthalpy will be large.The lattice enthalpies of both carbonates and oxides fall as you go down the Group because the positive ions are getting bigger. The inter-ionic distances are increasing and so the attractions become weaker.

	ionic radius (nm)
Mg^{2+}	0.065
Ca^{2+}	0.099
O^{2-}	0.140
CO_3^{2-}	?

The lattice enthalpies fall at different rates because of the different sizes of the two negative ions - oxide and carbonate. The oxide ion is relatively small for a negative ion (0.140 nm), whereas the carbonate ion is large.In the oxides, when you go from magnesium oxide to calcium oxide, for example, the inter-ionic distance increases from 0.205 nm (0.140 + 0.065) to 0.239 nm (0.140 + 0.099) - an increase of about 17%.In the carbonates, the inter-ionic distance is dominated by the much larger carbonate ion. Although the inter-ionic distance will increase by the same amount as you go from magnesium carbonate to calcium carbonate, as a percentage of the total distance the increase will be much less.The carbonate ion radius was 0.3 nm. The inter-ionic distances in the two cases we are talking about would increase from 0.365 nm to 0.399 nm - an increase of only about 9%.The rates at which the two lattice energies fall as you go down the Group depends on the percentage change as you go from one compound to the next. On that basis, the oxide lattice enthalpies are bound to fall faster than those of the carbonates.

Nitrates

The nitrate ion is bigger than an oxide ion, and so its radius tends to dominate the inter-ionic distance. The lattice enthalpy of the oxide will again fall faster than the nitrate.

Beryllium Chloride is Covalent

Beryllium chloride, $BeCl_2$, melts at 405°C and boils at 520°C. That compares with 714°C and 1412°C for magnesium chloride.Notice how much dramatically lower the boiling point of beryllium chloride is compared with magnesium chloride. The much higher boiling point of magnesium chloride is what you might expect from the strong forces between the positive and negative ions present.Because its boiling point is much lower, it follows that beryllium chloride can't contain ions - it must be covalent. On the other hand, the melting point is quite high for such a small covalent molecule.

Reaction with Water

Beryllium chloride reacts vigorously and exothermically with water with the evolution of acidic, steamy hydrogen chloride gas. This is typical of covalent chlorides.In the first instance, it reacts to give hydrated beryllium ions, $[Be(H_2O)_4]^{2+}$, and chloride ions.But the hydrated beryllium ions (called tetraaquaberyllium ions) are quite strongly acidic. The small beryllium ion at the centre attracts the electrons in the bonds towards itself, and that makes the hydrogen atoms in the water even more positive than they usually are. If the solution is hot and concentrated (as it is likely to be if you add water to solid beryllium chloride - a very exothermic reaction), chloride ions can remove one or more of these hydrogen ions to produce hydrogen chloride gas.All the other ionic chlorides in Group 2 dissolve in water without any obvious reaction.

The Structure of Beryllium Chloride

Beryllium chloride, $BeCl_2$, is a linear molecule with all three atoms in a straight line. Showing only the outer electrons:

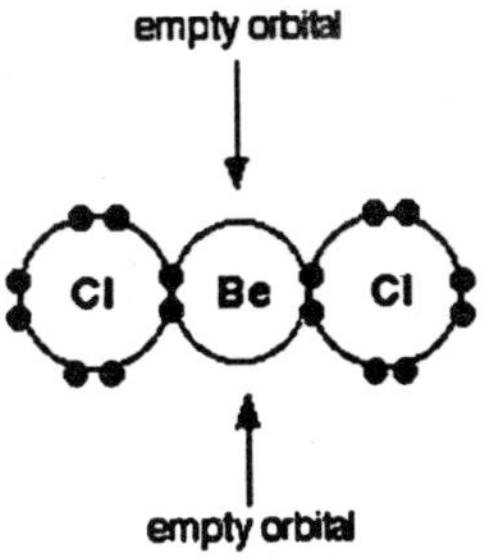

Beryllium chloride is known as an *electron-deficient compound* because it has the two empty orbitals at the bonding level.If it had this same simple structure as a solid, you would expect the melting point to be much lower than it actually is. It is a very small molecule, and so the intermolecular attractions would be expected to be fairly weak.In the solid, the $BeCl_2$ molecules polymerise to make long chains. They do this by forming coordinate bonds (dative covalent bonds) between lone pairs on chlorine atoms and adjacent beryllium atoms. The diagram shows a simple dimer - the start of the polymerisation process.

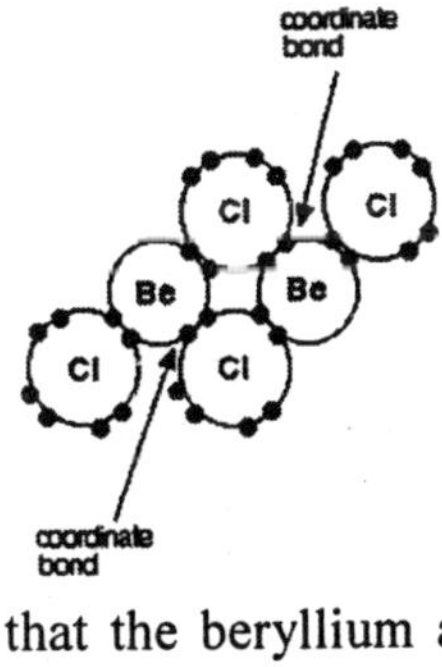

The last diagram that the beryllium atoms are still electron deficient. The process continues. The next diagram shows the coordinate bonds in the conventional way using arrows. The arrow goes from the atom which is supplying the pair of electrons to the atom with the empty orbital.

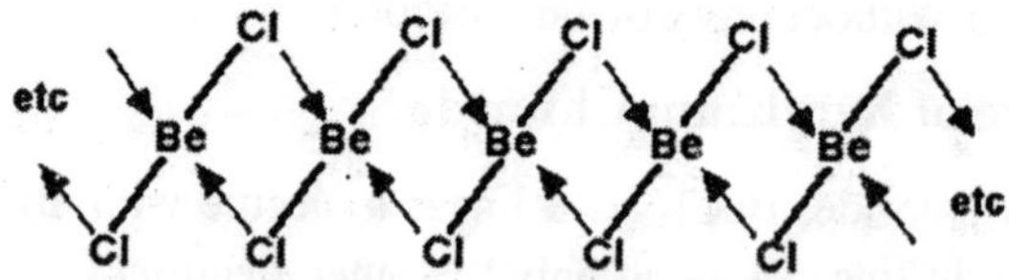

Make sure that you fully understand how this diagram relates to the dimer shown in the previous diagram.Beryllium has quite a high electronegativity compared with the rest of the Group. That means that it attracts a bonding pair of electrons towards itself more strongly than magnesium and the rest do.In order for an ionic bond to form, the beryllium has to let go of its electrons. It is too electronegative to do that.

Beryllium Forms 4-coordinated Complex Ions

Although beryllium doesn't normally form simple ions, Be^{2+}, it does form ions in solution. In these, the beryllium ion becomes attached to four water molecules to give a complex ion with the formula $[Be(H_2O)_4]^{2+}$.The ion is said to be *4-coordinated*, or to have a *coordination number of 4*, because there are four water molecules arranged around the central beryllium.Many hydrated metal ions are 6-coordinated. For example, magnesium ions in solution exist as $[Mg(H_2O)_6]^{2+}$.The water molecules in these ions are attached to the central metal ion via coordinate bonds (dative covalent bonds). One of the lone pairs on each water molecule is used to form a bond with an empty orbital in the metal ion.Each time one of these bonds is formed, energy is released, and the ion becomes more stable.

The Hydration of Beryllium

The problem is that there has to be somewhere that the lone pairs on the water molecules can attach to. Beryllium has the electronic structure $1s^2 2s^2$. It is helpful to draw this as an "electrons-in-boxes" diagram:

When beryllium forms a 2+ ion it loses the 2 electrons in the 2s orbital. That leaves the 2-level completely empty.

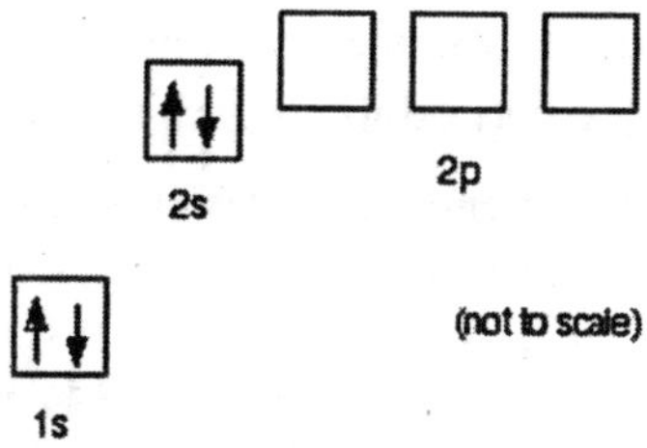

The 2-level orbitals reorganise themselves (hybridise) to make four equal orbitals, each of which can accept a lone pair of electrons from a water molecule. In the next diagram the 1s electrons have been left out. They aren't relevant to the bonding.

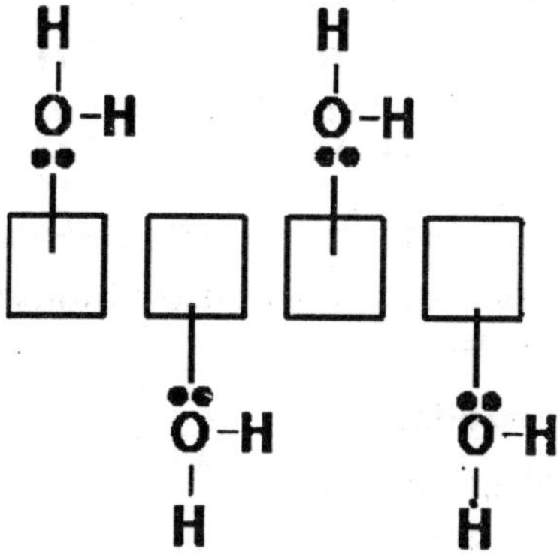

Each water molecule, of course, has two lone pairs of electrons. Only one of them is shown to avoid cluttering the diagram.

Notice that once four water molecules have bonded in this way, there isn't any more space available at the bonding level. All the empty orbitals from the original beryllium ion are being used.The water molecules arrange themselves to get as far apart as possible - which is pointing towards the corners of a tetrahedron. The ion therefore has a tetrahedral shape.

The Hydration of Magnesium

You might think that magnesium would behave just the same, but at the 3-level there are 3d orbitals available as well as 3s and 3p.When the magnesium ion is formed, it leaves empty 3s, 3p and 3d orbitals. When that ion is hydrated, it uses the 3s orbital, all three of the 3p orbitals and two of the 3d orbitals. These are reorganised to leave a total of six empty orbitals which are then used for bonding.

As the ions get bigger, there is less tendency for them to form proper coordinate bonds with water molecules. The ions become so big that they aren't sufficiently attractive to the lone pairs on the water molecules to form formal bonds - instead the water molecules tend to cluster more loosely around the positive ions. Where they do form coordinate bonds with the water, however, they will be 6-coordinated just like the magnesium.

Beryllium hydroxide is amphoteric

Amphoteric means that it can react with both acids and bases to form salts.

The other hydroxides of the Group 2 metals are all basic. They react with acids to form salts. For example:

$$Ca(OH)_{2(s)} + 2HCl_{(aq)} \longrightarrow CaCl_{2(aq)} + 2H_2O_{(l)}$$

Calcium hydroxide reacts with dilute hydrochloric acid to give calcium chloride and water.

Beryllium hydroxide

Beryllium hydroxide reacts with acids, forming solutions of beryllium salts. For example:

$$Be(OH)_{2(s)} + H_2SO_{4(aq)} \longrightarrow BeSO_{4(aq)} + 2H_2O_{(l)}$$

But it also reacts with bases such as sodium hydroxide solution. Beryllium hydroxide reacts with the sodium hydroxide to give a colourless solution of sodium tetrahydroxoberyllate.

$$Be(OH)_{2(s)} + 2NaOH_{(aq)} \longrightarrow Na_2Be(OH)_{4(aq)}$$

This contains the complex ion, $[Be(OH)_4]^{2-}$. The name describes this ion. Tetra means four; hydroxo refers to the OH groups; beryllate shows that the beryllium is present in a negative ion. The "ate" ending always shows that the ion is negative. The beryllium ions in solution exist as the hydrated ion, $[Be(H_2O)_4]^{2+}$. The beryllium has such a strongly polarising effect on the water molecules that hydrogen ions are very easily removed from them. The sodium hydroxide solution contains hydroxide ions which are powerful bases. If you add just the right amount of sodium hydroxide solution, you get a precipitate of what is normally called "beryllium hydroxide" - but which is a shade more complicated than that!

$$[Be(H_2O)_4]^{2+}(ag)+2OH^-(aq) \longrightarrow Be(H_2O)_2(OH)_2(aq)+2H_2O(l)$$

The product (other than water) is a neutral complex, and it is covalently bonded. All that has happened to the original complex ion is that two hydrogen ions have been removed from the water molecules.You get a precipitate of the neutral complex because of the lack of charge on it. There isn't enough attraction between this neutral complex and water molecules to bring it into solution.

The hydrogen ions that were originally removed are simply replaced. The precipitate dissolves as the original hydrated beryllium ion is re-formed.

$$Be(H_2O)_2(OH)_{2(s)} + 2H^+_{(aq)} \longrightarrow [Be(H_2O)_4]^{2+}_{(aq)}$$

Adding more hydroxide ions to the neutral complex pulls more hydrogen ions off the water molecules to give the tetrahydroxoberyllate ion:

$$Be(H_2O)_2(OH)_2(g)+2OH^-(aq) \longrightarrow [Be(OH)_4]^{2-}(aq)+2H_2O(l)$$

The beryllium hydroxide dissolves because the neutral complex is converted into an ion which will be sufficiently attracted to water molecules.

Calcium hydroxide is truly ionic - and contains simple hydroxide ions, OH^-. These react with hydrogen ions from an acid to form water - and so the hydroxide reacts with acids.There isn't any equivalent to the neutral complex. Adding more hydroxide ions from a base has no effect because they haven't got anything to react with.

Chapter 2

Structure of Atoms and Molecules

THE NATURE OF THE PROBLEM

The understanding and prediction of the properties of matter at the atomic level represents one of the great achievements of twentieth-century science. The theory developed to describe the behaviour of electrons, atoms and molecules differs radically from familiar Newtonian physics, the physics governing the motions of macroscopic bodies and the physical events of our everyday experiences. The discovery and formulation of the fundamental concepts of atomic physics in the period 1901 to 1926 by such men as Planck, Einstein, de Broglie and Heisenberg caused what can only be described as a revolution in the then-accepted basic concepts of physics.

The new theory is called *quantum theory* or *quantum mechanics*. As far as we now know this theory is able to account for all observable behaviour of matter and, with suitable extensions, for the interaction of matter with light. The proper formulation of quantum mechanics and its application to a specific problem requires a rather elaborate mathematical framework, as do proper statements and applications of Newtonian physics. We may, however, in this introductory account acquaint ourselves with the critical experiments which led to the formulation of quantum mechanics and apply the basic concepts of this new mechanics to the study of electrons.

Specifically the problem we set ourselves is to discover the physical laws governing the behaviour of electrons and then apply these laws to determine how the electrons are arranged when bound to nuclei to form atoms and molecules. This arrangement of electrons is termed the *electronic structure* of the atom or molecule. Furthermore, we shall discuss the relationship between the electronic structure of an atom and its physical properties, and how the electronic structure is changed during a chemical reaction.

Rutherford's nuclear model for the atom set the stage for the understanding of the structure of atoms and the forces holding them together.

From Rutherford's alpha-scattering experiments it was clear that the atom consisted of a positively-charged nucleus with negatively-charged electrons arranged in some fashion around it, the electrons occupying a volume of space many times larger than that occupied by the nucleus. (The diameters of nuclei fall in the range of $1 \times 10^{-12} \rightarrow 1 \times 10^{-13}$ cm, while the diameter of an atom is typically of the order of magnitude of 1×10^{-8} cm.) The forces responsible for binding the atom, and in fact all matter (aside from the nuclei themselves), are electrostatic in origin: the positively-charged nucleus attracts the negatively-charged electrons. There are attendant magnetic forces which arise from the motions of the charged particles. These magnetic forces give rise to many important physical phenomena, but they are smaller in magnitude than are the electrostatic forces and they are not responsible for the binding found in matter.

During a chemical reaction only the number and arrangement of the electrons are changed, the nucleus remaining unaltered. The unchanging charge of the atomic nucleus is responsible for retaining the atom's chemical identity through any chemical reaction. Thus for the purpose of understanding the chemical properties and behaviour of atoms, the nucleus may be regarded as simply a point charge of constant magnitude for a given element, giving rise to a central field of force which binds the electrons to the atom.

Rutherford proposed his nuclear model of the atom in 1911, some fifteen years before the formulation of quantum mechanics.

Consequently his model, when first proposed, posed a dilemma for classical physics. The nuclear model, based as it was on experimental observations, had to be essentially correct, yet all attempts to account for the stability of such a system using Newtonian mechanics ended in failure.

According to Newtonian mechanics we should be able to obtain a complete solution to the problem of the electronic structure of atoms once the nature of the force between the nucleus and the electron is known. The electrostatic force operative in the atom is well understood and is described by Coulomb's law, which states that the force between two particles with charges e_1 and e_2 separated by a distance R is given by:

$$F = \frac{e_1 e_2}{R^2}$$

There is a theorem of electrostatics which states that no stationary arrangement of charged particles can ever be in electrostatic equilibrium, i.e., be stable to any further change in their position. This means that all the particles in a collection of postively and negatively charged species will always have resultant forces of attraction or repulsion acting on them no matter how they are arranged in space. Thus no model of the atom which invokes some stationary arrangement of the electrons around the nucleus is possible. The electrons must be in motion if electrostatic stability is to be preserved. However, an electron moving in the field of a nucleus experiences a force and, according to Newton's second law of motion, would be accelerated. The laws of electrodynamics state that an accelerated charged particle should emit light and thus continuously lose energy. In this dynamical model of the atom, all of the electrons would spiral into the nucleus with the emission of light and all matter would collapse to a much smaller volume, the volume occupied by the nuclei.

No one was able to devise a theoretical model based on Newtonian, or what is now called classical mechanics, which would explain the electrostatic stability of atoms. The inescapable conclusion was that the classical equations of motion did not apply to the electron. Indeed, by the early 1900's a number of physical phenomena dealing with light and with events on the atomic level were found to be inexplicable in terms of classical mechanics. It became increasingly clear that Newtonian mechanics, while predicting with precision the motions of masses ranging in size from stars to microscopic particles, could not predict the behaviour of particles of the extremely small masses encountered in the atomic domain. The need for a new set of laws was indicated.

SOME IMPORTANT EXPERIMENTS WITH ELECTRONS AND LIGHT

Certainly the early experiments on the properties of electrons did not suggest that any unusual behaviour was to be expected. Everything pointed to the electron being a particle of very small mass. The trajectory of the electron can be followed in a device such as a Wilson cloud chamber. Similarly, a beam of electrons generated by passing a current between two electrodes in a glass tube from which the air has been partially evacuated will cast the shadow of an obstacle placed in the path of the beam. Finally, the particle nature of the electron was further evidenced by the determination of its mass and charge.

Just as classical considerations placed electrons in the realm of particles, the same classical considerations placed light in the realm of waves with equal certainty. How can one explain diffraction effects without invoking wave motion?

In the years from 1905 to 1928 a number of experiments were performed which could be *interpreted by classical mechanics* only if it was assumed that electrons possessed a wave motion, and light was composed of a stream of particles! Such dualistic descriptions, ascribing both wave and particle characteristics to electrons or light, are impossible in a physical sense. The electron must behave either as a particle or a wave, but not both (assuming it is either). "Particle" and "wave" are both concepts used by ordinary or classical mechanics and we see the paradox which results when classical concepts are used in an attempt to describe events on an atomic scale. We shall consider just a few of the important experiments which gave rise to the classical explanation of dual behaviour for the description of electrons and light, a description which must ultimately be abandoned.

The Photoelectric Effect

1. The number of electrons released from the surface increases as the intensity of he light is increased, but the energies of the emitted electrons are independent of the intensity of the light.
2. No electrons are emitted from the surface of the metal unless the frequency of the light is greater than a certain minimum value. When electrons are ejected from the surface they exhibit a range of velocities, from zero up to some maximum value. The energy of the electrons with the maximum velocity is found to increase linearly with an increase in the frequency of the incident light.

Certain metals emit electrons when they are exposed to a source of light. This is called the photoelectric effect. The pertinent results of this experiment are

The first result shows that light cannot be a wave motion in the classical sense. As an analogy, consider waves of water striking a beach and hitting a ball (in place of an electron) at the water's edge. The intensity of a wave is proportional to the square of the amplitude (or height) of the wave. Certainly, even when the frequency with which the waves strike the beach remains constant, an increase in the amplitude of the waves will cause much more energy to be released when they strike the beach and hit the ball. Yet when light "waves" strike a substance only the number of emitted electrons increases as the intensity is increased; the energy of the most energetic electrons remains constant. This can be explained only if it is assumed that the energy in a beam of light is not transmitted in the manner characteristic of a wave, but rather that the energy comes in bundles or packets and that the size of the packet is determined by the frequency of the light. This explanation put forward by Einstein in 1905 relates the energy to the frequency—and *not* to the intensity of the light—as required by the experimental results. A packet of light energy is called a photon. The results of the photoelectric experiment show that the energy ε of a photon is directly proportional to the frequency ν of the light, or, calling the constant of proportionality h, we have:

$$\varepsilon = h\nu$$

Since the electron is bound to the surface of the metal, the photon must possess a certain minimum amount of energy, i.e., possess a certain minimum frequency ν_0, just sufficient to free the electron from the metal. When an electron is ejected from the surface by a photon with a frequency greater than this minimum value, the energy of the photon in excess of the minimum amount appears as kinetic energy of the electron. Thus:

$$\text{Kinetic energy of electron} = h\nu - h\nu_0$$

where $h\nu$ is the energy of the photon with frequency ν, and $h\nu_0$ is the energy of the photon which is just sufficient to free the electron from the metal. Experimentally we can measure the kinetic energy of the electrons as a function of the frequency ν. A plot of the kinetic energy versus the frequency gives a straight line whose slope is equal to the value of h, the proportionality constant. The value of h is found to be 6.6×10^{-27} erg sec.

Equation is revolutionary. It states that the energy of a given frequency of light cannot be varied continuously, as would be the case classically, but rather that it is fixed and comes in packets of a discrete size. The energy of light is said to be quantized and a photon is one quantum (or bundle) of energy.

It is tempting at this point, if we desire a classical picture of what is happening, to consider each bundle of light energy, that is, each photon, to be an actual particle. Then one photon, on striking an individual electron, scatters the electron from the surface of the metal. The energy originally in the photon is converted into the kinetic energy of the electron (minus the energy required for the electron to escape from the surface). This picture must not be taken literally, for then the diffraction of light is inexplicable. Nor, however, can the wave picture for diffraction be taken literally, for then the photoelectric effect is left unexplained. In other words, light behaves in a different way from ordinary particles and waves and requires a special description.

The constant h determines the size of the light quantum. It is termed Planck's constant in honour of the man who first postulated that energy is not a continuously variable quantity, but occurs only in packets of a discrete size. Planck proposed this postulate in 1901 as a result of a study of the manner in which energy is distributed as a function of the frequency of the light emitted by an incandescent body. Planck was forced to assume that the energies of the oscillations of the electrons in the incandescent matter, which are responsible for the emission of the light, were quantized. Only in this way could he provide a theoretical explanation of the experimental results. There was a great reluctance on the part of scientists at that time to believe that Planck's revolutionary postulate was anything more than a mathematical device, or that it represented a result of general applicability in atomic physics. Einstein's discovery that Planck's hypothesis provided an explanation of the photoelectric effect as well indicated that the quantization of energy was indeed a concept of great physical significance. Further examples of the quantization of energy were soon forthcoming, some of which are discussed below.

The Diffraction of Electrons

Just as we have found dualistic properties for light when its properties are considered in terms of classical mechanics, so we find the same dualism for electrons. From the early experiments on electrons it was concluded that they were particles. However, a beam of electrons, when passed through a suitable grating, gives a diffraction pattern entirely analogous to that obtained in diffraction experiments with light. In other words, not only do electrons and light both appear to behave in completely different and strange ways when considered in terms of our everyday physics, they both appear to behave in the same way! Actually, the same strange behaviour can be observed for protons and neutrons. All the fundamental particles and light exhibit behaviour which leads to conflicting conclusions when classical mechanics is used to interpret the experimental findings.

The diffraction experiment with electrons was carried out at the suggestion of de Broglie. In 1923 de Broglie reasoned that a relationship should exist between the "particle" and "wave" properties for light. If light is a stream of particles, they must possess momentum. He applied to the energy of the photon Einstein's equation for the equivalence between mass and energy:

$$\varepsilon = mc^2$$

where c is the velocity of light and m is the mass of the photon. Thus the momentum of the photon is mc and:

$$\varepsilon = \text{momentum} \times c$$

If light is a wave motion, then of course it possesses a characteristic frequency ν and wavelength λ which are related by the equation:

$$\nu = \frac{c}{\lambda}$$

The frequency and wavelength may be related to the energy of the photon by using Einstein's famous relationship:

$$\varepsilon = h\nu = \frac{hc}{\lambda}$$

By equating the two expressions for the energy:

$$\frac{hc}{\lambda} = \text{moment} \times c$$

de Broglie obtained the following relationship which bears his name:

$$\lambda = h/\text{momentum}$$

However, de Broglie did not stop here. It was he who reasoned that light and electrons might behave in the same way. Thus a beam of electrons, each of mass m and with a velocity υ (and hence a momentum $m\upsilon$) should exhibit diffraction effects with an apparent wavelength:

$$\lambda = \frac{h}{m\upsilon}$$

Using de Broglie's relationship, we can calculate that an electron with a velocity of 1×10^9 cm/sec should have a wavelength of approximately 1×10^{-8} cm. This is just the order of magnitude of the spacings between atoms in a crystal lattice. Thus a crystal can be used as a diffraction grating for electrons. In 1927 Davisson and Germer carried out this very experiment and verified de Broglie's prediction.

Line Spectra

A gas will emit light when an electrical discharge is passed through it. The light may be produced by applying a large voltage across a glass tube containing a gas at a low pressure and fitted with electrodes at each end. A neon sign is an example of such a "discharge tube." The electrons flowing through the tube transfer some of their energy to the electrons of the gaseous atoms. When the atomic electrons lose this extra energy and return to their normal state in the atom the excess energy is emitted in the form of light. Thus the gaseous atoms serve to transform electrical energy into the energy of light. The puzzling feature of the emitted light is that when it is passed through a diffraction grating (or a prism) to separate the light according to its wavelength, only certain wavelengths appear in the spectrum. Each wavelength appears in the spectrum as a single narrow line of coloured light, the line resulting from the fact that the emitted light is passed through a narrow slit (thus producing a thin "line" of light) before striking the grating or the prism and being diffracted. Thus a "line" spectrum rather than a continuous spectrum is obtained when atomic electrons are excited by an electrical discharge.

An example of such a spectrum is given in Fig, which illustrates the visible spectrum observed for the hydrogen atom. This spectrum should be contrasted with the more usual continuous spectrum obtained from a source of white light which consists of a continuous band of colours ranging from red at the long wavelength end to violet at short wavelengths.

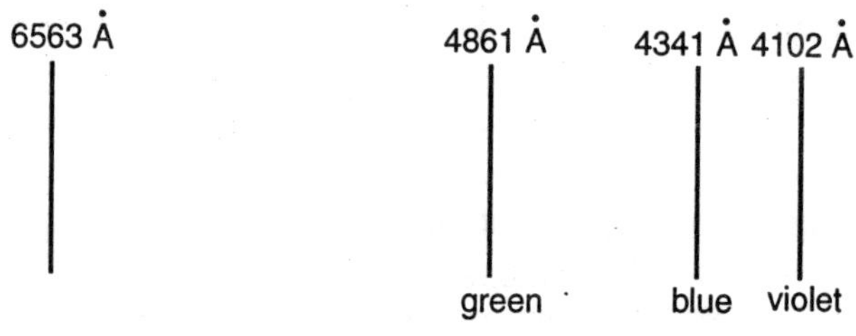

Fig. The visible spectrum for hydrogen atoms
(1Å = 1 Ångstrom = 1×10^{-8} cm)

The energy lost by an electron as it is attracted by the nucleus appears in the form of light. If all energies were possible for an electron when bound to an atom, all wavelengths or frequencies should appear in its emission spectrum, i.e., a continuous spectrum should be observed. The fact that only certain lines appear implies that only certain values for the energy of the electron are possible or allowed. We could describe this by assuming that the energy of an electron bound to an atom is quantized. The electron can then lose energy only in fixed amounts corresponding to the difference in value between two of the allowed or quantized energy values of the atom. Since the energy of a photon is given by

$$\varepsilon = h\nu$$

and ε must correspond to the difference between two of the allowed energy values for the electron, say E and E' $\{E' > E)$, then the value of the corresponding frequency for the photon will be given by

$$\frac{E'-E}{h} = \nu = \frac{\varepsilon}{h}$$

Obviously, if only certain values of E are allowed, only certain values of ε or ν will be observed, and a line spectrum rather than a continuous spectrum (which contains all values of ν) will be observed.

Equation was put forward by Bohr in 1913 and is known as Bohr's frequency condition. It was Bohr who first suggested that atomic line spectra could be accounted for if we assume that the energy of the electron bound to an atom is quantized. Thus the parallelism between the properties of light and electrons is complete. Both exhibit the wave-particle dualism and the energies of both are quantized.

The Compton Effect

The results of one more experiment will play an important role in our discussions of the nature of electrons bound to an atom. The experiment concerns the direct interaction of a photon and an electron.

In order to determine the position of an object we must somehow "see" it. This is done by reflecting or scattering light from the object to the observer's eyes. However, when observing an object as small as the electron we must consider the interaction of an individual photon with an individual electron. It is found experimentally—and this is the Compton effect—that when a photon is scattered by an electron, the frequency of the emergent photon is lower than it was before the scattering. Since $\varepsilon = h\nu$, and ν is observed to decrease, some of the photon's energy has been transmitted to the electron. If the electron was initially free, the loss in the energy of the photon would appear as kinetic energy of the electron. From the law of conservation of energy,

$$h\nu - h\nu' = \frac{1}{2}m\upsilon^2$$

$$= \text{kinetic energy of electron}$$

where ν' is the frequency of the photon after collision with the electron. This experiment brings forth a very important effect in the making of observations on the atomic level. We cannot make an observation on an object without at the same time disturbing the object. Obviously, the electron receives a kick from the photon during the observation. While it is possible to determine the amount of energy given to the electron by measuring ν and ν', we cannot however, predict in advance the final momentum of the electron. A knowledge of the momentum requires a knowledge of the direction in which the electron is scattered after the collision and while this can be measured experimentally one cannot predict the outcome of any given encounter. We shall illustrate later, with the aid of a definite example, that information regarding both the position and the momentum of an electron cannot be obtained with unlimited accuracy. For the moment, all we wish to draw from this experiment is that we must be prepared to accept a degree of uncertainty in the events we observe on the atomic level. The interaction of the observer with the system he is observing can be ignored in classical mechanics where the masses are relatively large. This is not true on the atomic level as here the "tools" employed to make the observation necessarily have masses and energies comparable to those of the system we are observing.

In 1926 Schrodinger, inspired by the concept of de Broglie's "matter waves," formulated an equation whose role in solving problems in atomic physic's corresponds to that played by Newton's equation of motion in classical physics. This single equation will correctly predict all physical behaviour, including, for example, the experiments with electrons and light discussed above. Quantization follows automatically from this equation, now called Schrodinger's equation, and its solution yields all of the physical information which can be known about a given system. Schrodinger's equation forms the basis of quantum mechanics and as far as is known today the solutions to all of the problems of chemistry are contained within the framework of this new mechanics. We shall in the remainder of this site concern ourselves with the behaviour of electrons in atoms and molecules as predicted and interpreted by quantum mechanics.

UNITS OF MEASUREMENT USED IN ATOMIC PHYSICS

The energies of electrons are commonly measured and expressed in terms of a unit called an *electron volt.* An electron volt (ev) is defined as the energy acquired by an electron when it is accelerated through a potential difference of one volt.

Imagine an evacuated tube which contains two parallel separate metal plates connected externally to a battery supplying a voltage V. The cathode in this apparatus, the negatively-charged plate, is assumed to be a photoelectric emitter. Photons from an external light source with a frequency ν_o upon striking the cathode will supply the electrons with enough energy to just free them from the surface of the cathode. Once free, the electrons will be attracted by and accelerated towards the positively-charged anode. The electrons, which initially have zero velocity at the cathode surface, will be accelerated to some velocity υ when they reach the anode. Thus the electron acquires a kinetic energy equal to ½ $m\upsilon^2$ in falling through a potential of V volts. If the charge on the electron is denoted by e this same energy change in ev is given by the charge multiplied by the voltage V:

$$\frac{1}{2}m\upsilon^2 - eV$$

For a given velocity υ in cm/sec, equation provides a relationship between the energy unit in the cgs (centimetre, gram, second) system, the erg, and the electron volt. This relationship is:

The regular cgs system of units is inconvenient to use on the atomic level as the sizes of the standard units in this system are too large. Instead, a system of units called *atomic units*, based on atomic values for energy, length, etc., is employed.

Atomic units are defined in terms of Planck's constant and the mass and charge of the electron:

Planck's constant = $h = 6.625\times10^{-25}$ erg-sec

mass of electron = $m = 9.108\times10^{-28}$ g

charge on electron = $e = 4.8029\times10^{-10}$ esu

Length.

$$1au = a_0$$

$$= \frac{h^2}{4\pi^2 me^2}$$

$$= 0.52917\times10^{-18} cm$$

Force. Force has the dimensions of charge squared divided by distance squared or

$$1\text{au} = \frac{e^2}{a_0^2}$$

$$= 8.2377\times10^{-3}\text{dynes}$$

Energy. Energy is force acting through a distance or

$$1\text{au} = \frac{e^2}{a_0^2}$$

$$= 4.3592\times10^{-11}$$

$$= 2.7210\times10^{1}\text{ev}$$

THE NEW PHYSICS

Now that we have studied some of the properties of electrons and light and have seen that their behaviour cannot be described by classical mechanics, we shall introduce some of the important concepts of the new physics, quantum mechanics, which does predict their behaviour. For the study of chemistry, we are most interested in what the new mechanics has to say about the properties of electrons whose motions are in some manner confined, for example, the motion of an electron which is bound to an atom by the attractive force of the nucleus. An atom, even the hydrogen atom, is a relatively complicated system because it involves motion in three dimensions. We shall consider first an idealized problem in just one dimension, that of an electron whose motion is confined to a line of finite length. We shall state the results given by quantum mechanics for this simple system and contrast them with the results given by classical mechanics for a similar system, but for a particle of much larger mass. Later, we shall indicate the manner in which the quantum mechanical predictions are obtained for a system.

A CONTRAST OF THE OLD AND NEW PHYSICS

Consider an electron of mass $m = 9 \times 10^{-28}$ g which is confined to move on a line L cm in length. L is set equal to the approximate diameter of an atom, 1×10^{-8}cm = 1Å. Consider as well a system composed of a mass of 1*g* confined to move on a line, say 1 metre in length. We shall apply quantum mechanics to the first of these systems and classical mechanics to the second.

ENERGY

As either mass moves from one end of its line to the other, the potential energy (the energy which depends on the position of the mass) remains constant. We may set the potential energy equal to zero, and all the energy is then kinetic energy (energy of motion). When the electron reaches the end of the line, we shall assume that it is reflected by some force. Thus at the ends of the line the potential energy rises abruptly to a very large value, so large that the electron can never "break through." We can plot potential energy versus position x along the line Fig.

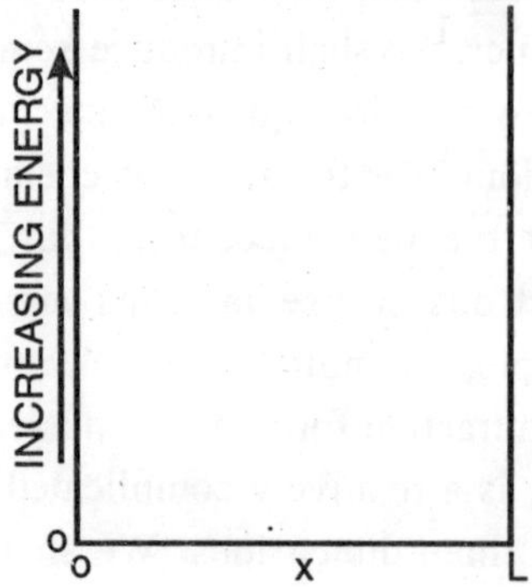

Fig. Potential energy diagram for a particle moving on a line of lenght L. When the electron is at $x = 0$ or $x = L$ The potential energy is infinite and for values of x between these limits ($0 < x < L$) the potential Energy is zero.

We refer to the electron (or the particle of $m = 1g$) as being in a potential well and we can imagine the abruptly rising potential at $x = 0$ and $x = L$ to be the result of placing a "wall" at each end of the line. First, what are the predictions of classical mechanics regarding the energy of the mass of 1 g? The total energy is kinetic energy and is simply:

$$E = KE = \frac{1}{2}m\upsilon^2$$

We know from experience that υ, the velocity, can have any possible value from zero up to very large values. Since all values for υ are allowed, all values for E are allowed. We conclude that the energy of a classical system may have any one of a continuous range of values and may be changed by any arbitrary amount. Let us contrast with this conclusion the prediction which quantum mechanics makes regarding the energy of an electron in a corresponding situation.

The quantum mechanical results are remarkable indeed, although they should not be surprising when we recall Bohr's explanation of the line spectra which are observed for atoms. Quantum mechanics predicts that there are only certain values of the energy which the electron confined to move on the line can possess. The energy of the electron is quantized. If this result could be observed for a massive particle, it would mean that only certain velocities were possible, say $\upsilon = 1$ cm/sec or 10 cm/sec but with no intermediate values! But then an electron is not really a particle. The expression for the allowed energies as given by quantum mechanics for this simple system is:

$$E_n = \frac{h^2 n^2}{8mL^2} \qquad n = 1, 2, 3, 4.......$$

where again h is Planck's constant and n is an integer which may assume any value from one to infinity. Since only discrete values for E are possible, the appearance of the index n in equation is necessary. A number such as n which appears in the expression for the energy is called a quantum number. Each value of the quantum number n fixes a value of E_n, one of the allowed energy values. We can indicate the possible values for the energy on an energy diagram. It is clear from equation that for given values of m and L, E_n equals a constant ($K = h^2/8mL^2$) multiplied by n^2:

$$E_n = Kn^2 \qquad n = 1, 2, 3, 4.......$$

Thus we can express the value of E_n in terms of so many units of K.

Each line, called an energy level, in Fig denotes an allowed energy for the system and the figure is called an energy level diagram. Each level is identified by its value of n as a subscript. A corresponding diagram for the case of the classical particle would consist of an infinite number of lines with infinitesimally small spacings between them, indicating that the energy in a classical system may vary in a continuous manner and may assume any value. The energy *continuum* of classical mechanics is replaced by a *discrete* set of energy levels in quantum mechanics.

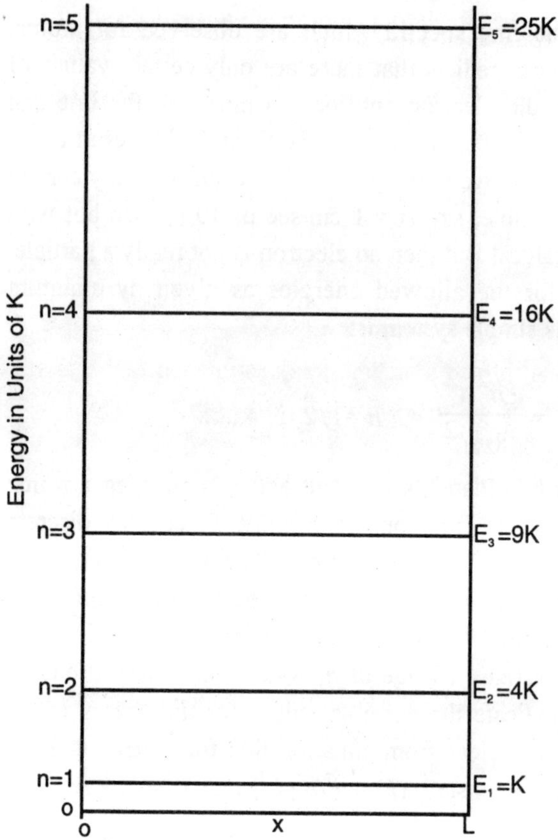

Fig. Energy level diagram for an electron moving on a line of length L. Only the first few levels are shown.

Suppose we could give the electron sufficient energy to place it in one of the higher (excited) energy levels. Then when it "fell" back down to the lowest value of E (called the ground level, E_1), a photon would be emitted. The energy ε of the photon would be given by the difference in the values of E_n and E_1 and, since $\varepsilon = h\nu$ the frequency of the photon would be given by the relationship:

$$\nu = \frac{E_n - E_1}{h} \qquad n = 1, 2, 3, 4, 5.......$$

which is Bohr's frequency condition (I-4). Thus only certain frequencies would be emitted and the spectrum would consist of a series of lines.

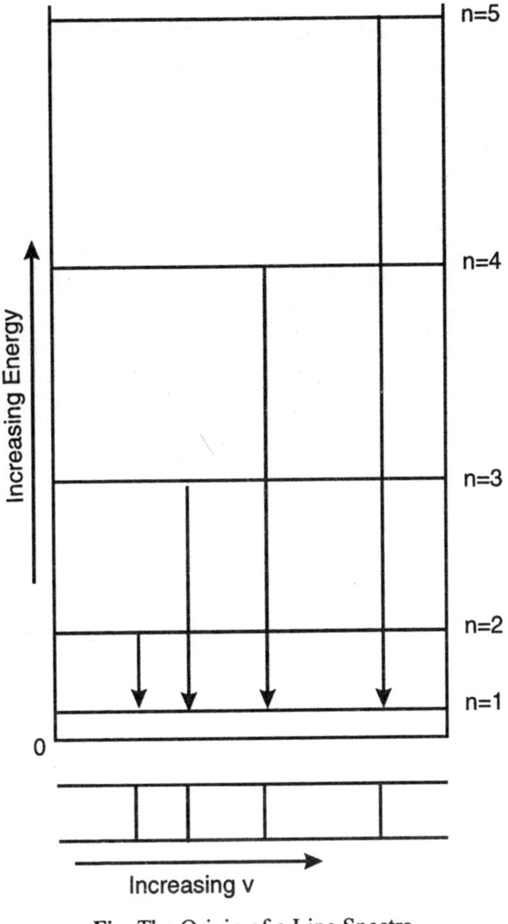

Fig. The Origin of a Line Spectra.

We can illustrate the change in energy when the electron falls to the lowest energy level by connecting the upper level and the $n = 1$ level by an arrow in an energy level diagram. The frequency of the photon emitted during the indicated drop in energy is proportional to the length of the arrow, i.e., to the change in energy. The line directly beneath each arrow represents the value of the frequency for that drop in energy. Since the differences in the lengths of the arrows increase as n increases, the separations between the observed frequencies show a corresponding increase. The spectrum, therefore, consists of a series of lines, with the spacings between the lines increasing as n increases. If the energy was not quantized and all values were possible, all jumps in energy would be possible and all frequencies would appear. Thus a continuum of possible energy values will produce a continuous spectrum of frequencies. A line spectrum, on the other hand, is a direct manifestation of the quantization of energy.

In the quantum case, as in the classical case, all of the energy will be in the form of kinetic energy. We may obtain an expression for the momentum of the electron by equating the total value of the energy E_n to $p^2/2m$, where p is the momentum ($= m\upsilon$) of the electron, ($p^2/2m$ is another way of expressing $2m\upsilon^2$.)

$$E_n = \frac{n^2h^2}{8mL^2}$$

$$= \frac{1}{2}m\upsilon^2$$

$$= \frac{p^2}{2m}$$

This gives:

$$p = \pm\frac{nh}{2L} \qquad n = 1, 2, 3, 4, 5.......$$

A plus and a minus sign must be placed in front of the number which gives the magnitude of the momentum to indicate that we do not know and cannot determine the direction of the motion. If the electron moves from left to right the sign will be positive. If it moves from right to left the sign will be negative. The most we can know about the momentum itself is its average value. This value will clearly be zero because of the equal probability for motion in either direction. The average value of p^2, however, is finite.

Since the lowest allowed value of the quantum number n in the quantum mechanical expression for the energy is unity, it is evident that the energy can never equal zero.

A confined electron can never be motionless. The expression for E_n also indicates that the kinetic energy and the momentum increase as the length of the line L is decreased. Thus the kinetic energy and momentum of the electron increase as its motion becomes more confined. This is both an important and a general result and will be referred to again.

POSITION

The concept of a trajectory is fundamental to classical mechanics. Given a particular mass with a given initial velocity and a knowledge of the forces acting on it, we may use classical mechanics to predict the exact position and velocity of the particle at any future time. Thus we speak of the trajectory of the particle and we may calculate it to any desired degree of accuracy. It is also possible, within the framework of classical mechanics, to measure the position and velocity of a particle at any given instant of time. Thus classical mechanics correctly predicts what one can experimentally measure for massive particles.

The difficulties which are encountered when we attempt to determine the position of an electron. The results of the Compton effect indicate that part of the energy of the photon used in making the observation is transferred to the electron, and we invariably disturb the electron when we attempt to measure its position. Thus it is not surprising to find that quantum mechanics does not predict the position of an electron exactly. Rather, it provides only a *probability* as to where the electron will be found. When we consider the experiments which attempt to define the position of. the electron, we shall find that this is the maximum information that can indeed be obtained even experimentally. The new mechanics again predicts only what can indeed be measured experimentally. We shall illustrate the probability aspect in terms of the system of an electron confined to motion along a line of length L. Quantum mechanical probabilities are expressed in terms of a distribution function which in this particular case we shall label $P_n(x)$.

Consider the line of length L to be divided into a large number of very small segments, each of length Δx. Then the probability that the electron is in one particular small segment Δx of the line is given by the product of Δx and the value of the probability distribution function $P_n(x)$ for that interval. For example, the probability distribution function for the electron when it is in the lowest energy level, $n = 1$, is given by $P_1(x)$.

The probability that the electron will be in the particular small interval Δx indicated in Fig. is equal to the shaded area, an area which in turn is equal to the product of Δx and the average value of $P_1(x)$ throughout the interval Δx, called $P_1(x)$,

$$\text{Probability that electron is in segement } \Delta x = P_1(x')\Delta x$$

The curve $P_1(x)$ may be determined in the following manner. We design an experiment able to determine whether or not the electron is in one particular segment Δx of the line when it is known to be in the quantum level $n = 1$. (One way in which this might be done is described below.) We perform the experiment a large number of times, say one hundred, for each segment and record the ratio of the number of times the electron is found in a particular segment to the total number of observations made for that segment. For example, an electron is found to be in the segment marked Δx (of length 0.1 L) in the figure for $P_1(x)$ in 18 out of 100 observations, or 18 per cent of the time. In the other 82 observations the electron was in one of the other segments. Thus the average value of $P_1(x)$ for this segment, called $P_1(x')$ must be $1.8/L$ since $P_1(x')\Delta x = (1.8/L)\ (0.1\ L) = 0.18$ or 18 per cent. A similar set of experiments is made for each of the segments Δx and in each case a rectangle is constructed with Δx as base and with a height equal to $P_1(x)$ such that the product $P_1(x)\Delta x$ equals the fractional number of times the electron is found in the segment Δx. The limiting case in which the total length L is divided into a very large number of very small segments ($\Delta x \rightarrow dx$) would result in the smooth curve shown in the figure for $P_1(x)$.

There is a different probability distribution for each value of E_n, or each quantum level, for example, by the probability distributions for the energy levels with $n = 2, 3, 4, 5$ and 6. The probability of finding the electron at the positions where the curve touches the x-axis is zero. Such a zero is termed a *node*. The number of nodes is always n-1 if we do not count the nodes at the ends of each $P_n(x)$ curve.

Let us first contrast these results, particularly that for $P_1(x)$, with the corresponding classical case. Since a classical analysis allows us to determine the position of a particle uniquely at any instant, either theoretically or experimentally, the idea of a probability distribution is foreign to a classical mechanical analysis. However, we still can determine the classical probability distribution for the particle confined to motion on a line. Since there are no forces acting on the particle as it traverses the line, it will be equally likely to be found at any point on the line. This probability will be the same regardless of the energy. There is again a striking difference between the classical and the quantum mechanics results. For the first quantum level, the graph of $P_1(x)$ indicates the electron will most likely be found at the midpoint of the line. Furthermore, the form of $P_n(x)$ changes with every change in energy. *Every allowed value of the of the energy has associated with it a distinct probability distribution for the electron.* Theses are the predictions of quantum mechanics regarding the position of a bound electron. Now let us investigate the experimental aspect of the problem to gain some physical reason for these predictions.

Let us design an experiment in which we attempt to pinpoint the position of an electron within a segment Δx. The experiment is a hypothetical one in that we imagine that we are to observe the electron through a microscope by reflecting or scattering light from it. Imagine the lens of a microscope being placed above the line L with the light entering from the side. The electron, when illuminated with light, will act as a small source of light and will produce at A an image in the form of a bright disc surrounded by a group of rings of decreasing intensity. Because of this effect, which is entirely analogous to the diffraction effect observed for a pinhole source of light, the centre of the image will appear bright even if the electron is not precisely located at the point marked x. It could equally well have been at any value of x between the points x' and x" and produced an image visible to the eye at A if the difference in the path lengths Bx' and Cx' (or Bx" and Cx") is less than one half of a wavelength. In other words the resolving power of a microscope is not unlimited but is instead determined by the wavelength of the light used in making the observation. The use of the microscope imposes an inherent uncertainty in our observation of the position of the electron. With the condition that the difference in the path lengths to the outside rim of the lens must be no greater than one half a wavelength and with the use of some geometry, the magnitude of the uncertainty in the position of the electron, $x'' - x' = \Delta x$, is found to be given approximately by:

$$\Delta x \sim \frac{\lambda}{2\sin\theta}$$

where θ is the angle indicated in the diagram.

Remembering the Compton effect and bearing in mind that we wish to disturb the electron as little as possible during the observation, we shall inquire as to the results obtained when a single photon is scattered from the electron. A single photon will not yield the complete diffraction pattern at *A*, but will instead produce a single flash of light. A diffraction pattern is the result of many photons passing through the microscope and represents the probability distribution for the emergent photons when they have been scattered by an electron lying between x' and x''. A single photon, when scattered from an electron within the length Δx, is however still diffracted and will produce a flash of light somewhere in one of the areas defined by the probability distribution produced by many photons passing through the system.

Thus even when we use but a single photon in our apparatus the uncertainty Δx in our experimentally determined position of the electron will still be given by equation. Obviously, if we want to locate an electron which is confined to move on a line to within a length that is small compared to the length of the line, we must use light which has a wavelength much less than *L*. This is exactly what equation states: the shorter the wavelength of the light which is used to observe the electron, the smaller will be the uncertainty Δx. That being the case, why not do the experiment with light of very short wavelength compared to the length *L*, say $\lambda = (1/100)L$? Then we can hope to find the electron on one small segment of the line, each segment being approximately $(1/100)L$ in length. Let us calculate the frequency and energy of a photon which has the required wavelength of $\lambda = (1/100)L$. As before, we set *L* equal to a typical atomic dimension of 1×10^{-8} cm.

$$\begin{aligned}\varepsilon &= h\nu \\ &= \frac{hc}{\lambda} \\ &= \frac{6.6\times10^{-27}\times3\times10^{10}}{10^{-10}}\end{aligned}$$

We are immediately in difficulty, because the energy of the electron in the first quantum level is easily found to be:

$$\begin{aligned}E_1 &= \frac{\left(6.6\times10^{-27}\right)^2}{9.1\times10^{-28}\times8\times10^{-16}} \\ &= 6.0\times10^{-11}\text{ergs} \\ &= K\end{aligned}$$

The energy of the photon is approximately 1×10^4 times greater than the energy of the electron! We know from the Compton effect that the collision of a photon with an electron imparts energy to the electron. Thus the electron after the collision will certainly not be in the state $n = 1$. It will be excited to oneCwe don't know whichCof the excited levels with $n = 2$ ($E = 4K$) or $n = 3$ ($E = 9K$), etc. The result is clear. If we demand an intimate knowledge of what the position of the electron is in a given state, we can obtain this information only at the expense of imparting to the electron an unknown amount of energy which destroys the system, i.e., the electron is no longer in the $n = 1$ level but in one of the other excited levels. If this experiment was repeated a large number of times and a record kept of the number of times an electron was located in each segment of the line (roughly $(1/100)L$), a probability plot similar to Fig. would be obtained.

We can ask another kind of question regarding the position of the electron: "How much information can be obtained about the position of the electron in a given quantum level without at the same time destroying that level?" The electron cannot accept energy in an amount less than that necessary to excite it to the next quantum level, $n = 2$. The difference in energy between E_2, and E_1, is $3K$.

Thus if we are to leave the electron in a state of known energy and momentum we must use light whose photons possess an energy less than $3K$.

Let us calculate the wavelength of the light with $\varepsilon = 2K$ and compare this value with the length L.

$$\lambda = \frac{hc}{\varepsilon}$$

$$= \frac{6.6\times10^{-27}\times3.0\times10^{10}}{12\times10^{-11}}$$

$$= 1.7\times10^{-6}\ \text{cm}$$

The wavelength is greater than the length of the line L. From equation it is clear that the uncertainty in the position of the particle will be of the order of magnitude of, or greater than, L itself. The electron will appear to be blurred over the complete length of the line in a single experiment! Thus there are two interpretations which can be given to the probability distributions, depending on the experiment which is performed. The first is that of a true probability of finding the electron in a given small segment of the line using light of very short λ relative to L. This experiment excites the electron, changes the system and leaves the electron with an unknown amount of energy and momentum. We have destroyed the object of our investigation. We now know where it was in a given experiment but not where it will be, in terms of energy or position.

Alternatively, we could use light with a λ approximately equal to L. This does not excite the electron and leaves it in a known energy level. However, now the knowledge of the position is very uncertain. The photons are scattered from the system and give us directly the smeared distribution P_1 pictured in Fig. In a real sense we must accept the fact that when the electron remains in a given state it is "smeared out" and "looks like" the pictures given for P_n. Thus we can interpret the P_n's as instantaneous pictures of the electron when it is bound in a known state, and forgot their probability aspect. This "smeared out" distribution is given a special name; it is called the *electron density distribution*. There will be a certain fraction of the total electronic charge at each point on the line, and when we consider a system in three dimensions, there will be a certain fraction of the total electronic charge in every small volume of space. Hence it is given the name electron density, the amount of charge per unit volume of space. The P_n's represent a charge density distribution which is considered static as long as the electron remains in the *n*th quantum level. Thus the P_n functions tell us either (a) the *fraction of time* the electron is at each point on the line for observations employing light of short wavelength, or (b) they tell us the *fraction of the total charge* found at each point on the line (the whole of the charge being spread out) when the observations are made with light of relatively long wavelength.

The electron density distributions of atoms, molecules or ions in a crystal can be determined experimentally by X-ray scattering experiments since X-rays can be generated with wavelengths of the same order of magnitude as atomic diameters (1×10^{-8} cm). In X-ray scattering the intensity of the scattered beam and the angle through which it is scattered are measured. The distribution of negative charge within the crystal scatters the X-rays and determines the intensity and angle of scattering. Thus these experimental quantities can be used to calculate the form of the electron density distribution.

There is a definite quantum mechanical relationship governing the magnitudes of the uncertainties encountered in measurements on the atomic level. We can illustrate this relationship for the one-dimensional system. Let us consider the minimum uncertainty in our observations of the position and the momentum of the electron moving on a line obtained in an experiment which leaves the particle bound in a given quantum level, say $n = 1$. This will require the use of light with $\lambda \sim L$. We have seen that the use of light of this wavelength limits us to stating that the electron is somewhere on the line of length L. We can say no more than this with certainty unless we use light of much shorter λ, and then we will change the quantum number of the electron. The uncertainty in the value of the position coordinate, which we shall call Δx, is just L, the length of the line:

$$\Delta x = L$$

We have previously shown that the momentum of the electron in the nth quantum level is given by:

$$P_n = \pm\frac{mh}{2L}, \text{ where } n = 1, 2, 3, \ldots\ldots$$

the plus and minus signs denoting the fact that while we know the magnitude of the momentum we cannot determine whether the electron is moving from left to right ($+nh/2L$) or from right to left ($-nh/2L$). The minimum uncertainty in our knowledge of the momentum is the difference between these two possibilities, or for $n = 1$:

$$\Delta p = +\frac{h}{2L} - \left(\frac{-h}{2L}\right)$$

$$= \frac{h}{L}$$

The product of the uncertainties in the position and the momentum is:

$$\Delta p \Delta x = L\frac{h}{L}$$

$$= h$$

This result is a *particular example* of a general relationship governing the product of the uncertainties in the momentum and position known as *Heisenberg's uncertainty principle.* In the general case, the equality sign in the above equation is replaced by the symbol "≥" which denotes that the product in the uncertainties $\Delta p\Delta x$ equals or exceeds the value of Planck's contant h, that is, the general statement is given by $\Delta p\Delta x \geq h$.

If we endeavour to decrease the uncertainty in the position coordinate (i.e., make Δx small) there will be a corresponding increase in the uncertainty of the momentum of the electron along the same coordinate, such that the product of the two uncertainties is always equal to Planck's constant. We saw this effect in our experiments wherein we employed light of short λ to locate the position of the electron more precisely. When we did this we excited the electron to one of the other available quantum states, thus making a knowledge of the energy and hence the momentum uncertain. We might also try to defeat Heisenberg's uncertainty principle by decreasing the length of the line L. By shortening L, we would decrease the uncertainty as to where the electron is. However, as was noted previously, the momentum increases as L is decreased and the uncertainty in p is always the same order of magnitude as p itself; in this case twice the magnitude of p. Thus the decrease in Δx obtained by decreasing L is offset by the increase in Δp which accompanies the increased confinement of the electron; the product $\Delta x \Delta p$ remains unchanged in value.

We can illustrate the operation of Heisenberg's uncertainty principle for a free particle by referring again to our hypothetical experiment in which we attempted to locate the position of an electron by using a microscope. We imagine the electron to be free and travelling with a *known momentum* in the direction of the x-axis with a photon entering from below along the y-axis. When the photon is scattered by the electron it *may* transfer momentum to the electron and continue on a line which makes an angle θ' to the y-axis. The photon, in doing so, will acquire momentum in the direction of the x-axis, a direction in which it initially had none. Since momentum must be conserved, the electron will receive a recoil momentum, a momentum equal in magnitude but opposite in direction to that gained by the photon. This is the Compton effect. Thus our act of observing the electron will lead to an uncertainty in its momentum as the amount of momentum transferred during the collision is uncontrollable. We may, however, set limits on the amount transferred and in this way determine the uncertainty introduced into the value of the momentum of the electron.

The momentum of the photon before the collision is all directed along the y-axis and has a magnitude equal to h/λ. After colliding with the electron the photon may be scattered to the left or to the right of the y-axis through any angle θ' lying between 0 and θ and still be collected by the lens of the microscope and seen by the observer at A. Thus every photon which passes through the microscope will have an uncertainty of $2(h/\lambda)\sin\theta$ in its component of momentum along the x-axis since it *may* have been scattered by the maximum amount to the left and acquired a component of $-(h/\lambda)\sin\theta$ or, on the other hand, it *may* have been scattered by the maximum amount to the right and acquired a momentum component of $+(h/\lambda)\sin\theta$. Any x-component of momentum acquired by the photon must have been lost by the electron and the uncertainty introduced into the momentum of the electron by the observation is also equal to $2(h/\lambda)\sin\theta$.

In addition to the uncertainty induced in the momentum of the electron by the act of measurement, there is also an inherent uncertainty in its position because of the limited resolving power of the microscope. The product of the two uncertainties at the instant of measurement or immediately following it is:

$$\Delta p \Delta x \sim 2\frac{h}{\lambda}\sin\theta \times \frac{\lambda}{2\sin\theta} = h$$

Heisenberg's uncertainty relationship is again fulfilled. Our experiment employs only a single photon which, since light itself is quantized, represents the smallest packet of energy and momentum which we can use in making the observation. Even in this idealized experiment the act of observation creates an unavoidable disturbance in the system.

DEGENERACY

We may use an extension of our simple system to illustrate another important quantum mechanical result regarding energy levels. Suppose we allow the electron to move on the x–y plane rather than just along the x-axis. The motions along the x and y directions will be independent of one another and the total energy of the system will be given by the sum of the energy quantum for the motion along the x-axis plus the energy quantum for motion along the y-axis. Two quantum numbers will now be necessary, one to indicate the amount of energy along each coordinate. We shall label these as n_x and n_y. Let us assume that the motion is confined to a length L along each axis, then:

$$E_{n_x,n_y} = \frac{h^2}{8mL^2}n_x^2 + \frac{h^2}{8mL^2}n_y^2$$

$$= \frac{h^2}{8mL^2}\left(n_x^2 + n_y^2\right), \text{ where } n = 1, 2, 3, \ldots\ldots$$

Nothing new is encountered when the electron is in the lowest quantum level for which $n_x = n_y = 1$. The energy $E_{1,1}$ simply equals $2h^2/8mL^2$.

Since two dimensions (x and y) are now required to specify the position of the electron, the probability distribution $P_{1,1}(x,y)$ must be plotted in the third dimension. We may, however, still display $P_{1,1}(x,y)$ in a two-dimensional diagram in the form of a contour map. All points in the x–y plane having the same value for the probability distribution $P_{1,1}(x,y)$ are joined by a line, a contour line. The values of the contours increase from the outermost to the innermost, and the electron, when in the level $n_x = n_y = 1$, is therefore most likely to be found in the central region of the x-y plane.

A plot of $P_{1,1}(x,y)$ along either of the axes indicated in Fig. (one parallel to the x-axis at $y = L/2$ and the other parallel to the y-axis at $x = L/2$) is similar in appearance to that for $P_1(x)$ shown in Fig.. That is, for a fixed value of y, the contribution to $P_{1,1}(x,y)$ from the motion along the y-axis is constant and

$$P_{1,1}(x, L/2) = \text{costant} \times P_1(x)$$

Thus, aside from the constant factor, $P_1(x)$ provides a profile, or if $P_{1,1}(x,y)$ were displayed in three dimensions, a cross section of the contour map of $P_{1,1}(x)$. A contour map is a display of the probability or density distribution in a plane; a profile is a display of the density distribution along a line.

Now consider the possibility of $n_x = 1$ and $n_y = 2$. Then

$$E_{1,2} = \frac{5h^2}{8mL^2}$$

We could also have the situation in which $n_x = 2$ and $n_y = 1$. This does not change the value of the total energy,

$$E_{2,1} = E_{1,2} = \frac{5h^2}{8mL^2}$$

but the probability distributions are different, $P_{1,1}(x,y) \neq P_{2,1}(x,y)$. When $n_x = 1$ and $n_y = 2$, there must be a node on the y-axis, i.e., a zero probability of finding the electron at $y = L/2$. Thus a slice through $P_{1,2}(x\ y)$ at $x = L/2$ parallel to the y-axis must be similar to the figure for $P_2(x)$, while a slice parallel to the x-axis will still be similar to $P_1(x)$. Just the reverse is true for the case $n_x = 2$ and $n_y = 1$. In this case, whether or not we can distinguish experimentally between the x-axis and y-axis, there are two different arrangements for the distribution of the electron, both of which have the same energy. The energy level is said to be degenerate. *The degeneracy of an energy level is equal to the number of distinct probability distribution for the system, all of which belong to this same energy level.*

The concept of degeneracy in an energy level has important consequences in our study of the electronic structure of atoms.

PROBABILITY AMPLITUDES

In quantum mechanics, Newton's familiar equations of motion are replaced by Schrödinger's equation. We shall not discuss this equation in any detail, nor indeed even write it down, but one important aspect of it must be mentioned. When Newton's laws of motion are applied to a system, we obtain both the energy and an equation of motion. The equation of motion allows us to calculate the position or coordinates of the system at any instant of time. However, when Schrödinger's equation is solved for a given system we obtain the energy directly, but not the probability distribution functionCthe function which contains the information regarding the position of the particle. Instead, the solution of Schrödinger's equation gives only the amplitude of the probability distribution function along with the energy. The probability distribution itself is obtained by *squaring the probability amplitude*. Thus for every allowed value of the energy, we obtain one or more (the energy value may be degenerate) probability amplitudes.

The probability amplitudes are functions only of the positional coordinates of the system and are generally denoted by the Greek letter ψ (psi). For a bound system the amplitudes as well as the energies are determined by one or more quantum numbers. Thus for every E_n we have one or more ψ_n's and by squaring the ψ_n's we may obtain the corresponding P_n's.

Let us look at the forms of the amplitude functions for the simple system of an electron confined to motion on a line. For any system, ψ is simply some mathematical function of the positional coordinates. In the present problem which involves only a single coordinate x, the amplitude functions may be plotted versus the x-coordinate in the form of a graph. The functions ψ_n are particularly simple in this case as they are sin functions.

$$\psi_n(x) = \sqrt{2/L}\sin(n\pi x/L) \text{ where } n = 1, 2, 3, \ldots\ldots$$

when squared, yields the corresponding P_n curves shown previously. When $n = 1$,

$$\psi_1(x) = \sqrt{2/L}\sin(\pi x/L)$$

When $x = 0$, $\psi_1(0) = \sqrt{2/L}\sin(0) = 0$

When $x = L$, $\psi_1(L) = \sqrt{2/L}\sin(\pi) = 0$

When $x = L/2$, $\psi_1(L/2) = \sqrt{2/L}\sin(\pi/2) = \sqrt{2/L}$

Thus ψ equals zero at $x = 0$ and $x = L$ and is a maximum when $x = L/2$. When this function is squared, we obtain:

$$P_1(x) = (2/L)\sin^2(\pi x/L)$$

and the graph previously given for $P_1(x)$.

As illustrated previously in Fig., the value of $\psi_n^2(x)$ or $P_n(x)$ multiplied by Δx, $\psi_n^2(x)\Delta x$, or $P_n(x)\Delta x$, is the probability that the electron will be found in some particular small segment of the line Δx. The constant factor of $\sqrt{2/L}$ which appears in every $\psi_n(x)$ is to assure that when the value of $\psi_n^2(x)\Delta x$ is summed over each of the small segments Δx, the final value will equal unity. This implies that the probability that the electron is somewhere on the line is unity, i.e., a certainty. Thus the probability that the electron is in any one of the small segments Δx (the value of $\psi_n^2(x)\Delta x$ or $P_n(x)\Delta x$ evaluated at a value of x between 0 and L) is a fraction of unity, i.e., a probability less than one.

Each ψ_n must necessarily go to zero at each end of the line, since the probability of the electron not being on the line is zero. This is a physical condition which places a mathematical restraint on the ψ_n. Thus the only acceptable ψ_n 's are those which go to zero at each end of the line. A solution of the form shown in Fig is, therefore, not an acceptable one. Since there is but a single value of the energy for each of the possible ψ_n functions, it is clear that only certain discrete values of the energy will be allowed. The physical restraint of confining the motion to a finite length of line results in the quantization of the energy. Indeed, if the line is made infinitely long (the electron is then free and no longer bound), solutions for any value of *n*, integer or non-integer, are possible; correspondingly, all energies are permissible. Thus only the energies of bound systems are quantized.

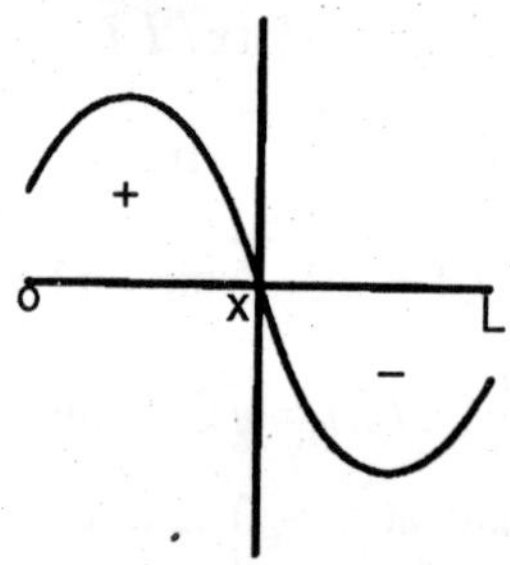

Fig. An Unacceptable form for $y_n(x)$.

The ψ_n 's have the appearance of a wave in that a given value of $\psi_n(x)$ is repeated as *x* is increased. They are periodic functions of *x*. We may, if we wish, refer to the wavelength of ψ_n. The wavelength of ψ_1 is $2L$ since only one half of a wave fits on the length *L*. The wavelength for ψ_2, is *L* since one complete wave fits in the length *L*. Similarly λ_3, = (2/3)L and λ_4 = (2/4)L. In general: $\lambda_n = 2L/n$, where n =1, 2, 3, ...

Because of the wave-like nature of the ψ_n 's, the new physics is sometimes referred to as wave mechanics, and the ψ_n functions are called wave functions. However, it must be stressed that a wave function itself has *no physical reality. All physical properties are determined by the product of the wave function with itself.* It is the *product* $\psi_n(x)\psi_n(x)$ which yields the physically measurable probability distribution. Thus ψ_n^2 may be observed but not ψ_n itself.

A ψ_n does not represent the trajectory or path followed by an electron in space. We have seen that the most we can say about the position of an electron is given by the probability function ψ_n^2. We do, however, refer to the wavelengths of electrons, neutrons, etc. But we must remember that the wavelengths refer only to a property of the amplitude functions and not to the motion of the particle itself.

A number of interesting properties can be related to the idea of the wavelengths associated with the wave functions or probability amplitude functions. The wavelengths for our simple system are given by $\lambda = 2L/n$. Can we identify these wavelengths with the wavelengths which de Broglie postulated for matter waves and which obeyed the relationship:

$$\lambda = h/p$$

The absolute value for the momentum (the magnitude of the momentum independent of its direction) of an electron on the line is $nh/2L$. Substituting this into de Broglie's relationship gives:

$$\begin{aligned} \lambda &= h/p \\ &= \frac{h}{nh/2L} \\ &= 2L/n \end{aligned}$$

So indeed the wavelengths postulated by de Broglie to be associated with the motions of particles are in reality the wavelengths of the probability amplitudes or wave functions. There is no need to postulate "matter waves" and the results of the electron diffraction experiment of Davisson and Germer for example can be interpreted entirely in terms of probabilities rather than in terms of "matter waves" with a wavelength $\lambda = h/p$.

It is clear that as n increases, λ becomes much less than L. For $n = 100$, ψ_{100} and P_{100} would appear as in Fig.. When $L >> \lambda_n$, the nodes in P_n are so close together that the function appears to be a continuous function of x. No experiment could in fact detect nodes which are so closely spaced, and any observation of the position of the electron would yield a result for P_{100} similar to that obtained in the classical case. This is a general result. When λ is smaller than the important physical dimensions of the system, quantum effects disappear and the system behaves in a classical fashion. This will always be true when the system possesses a large amount of energy, i.e, a high n value. When, however, λ is comparable to the physical dimensions of the system, quantum effects predominate.

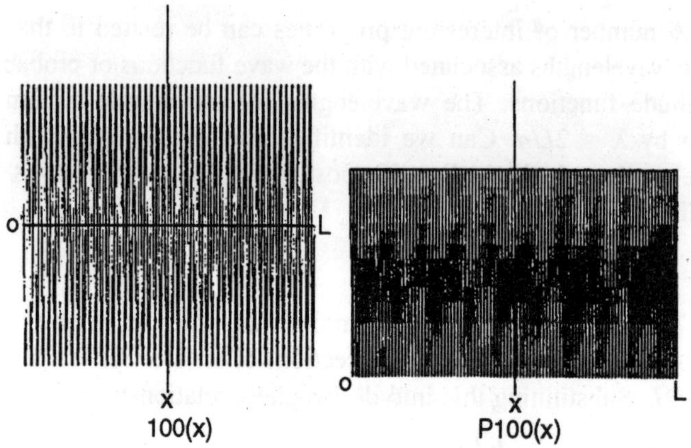

Fig. The Wave Function and Probability Distribution for $n = 100$.

Let us check to see whether or not quantum effects will be evident for electrons bound to nuclei to form atoms. A typical velocity of an electron bound to an atom is of the order of magnitude of 10^9 cm/sec. Thus:

$$\lambda = \frac{6.6\times10^{-27}}{9.1\times10^{-28}\times1\times10^{9}}$$

$$= 10^{-8}\,cm$$

This is a short wavelength, but it is of the same order of magnitude as an atomic diameter. Electrons bound to atoms will definitely exhibit quantum effects because the wavelength which determines their probability amplitude is of the same size as the important physical dimensionCthe diameter of the atom.

We can also determine the wavelength associated with the motion of the mass of 1 g moving on a line 1 m in length with a velocity of, say, 1 cm/sec:

$$\lambda = \frac{6.6\times10^{-27}}{1\times1}$$

$$= 6.6\times10^{-27}\,cm$$

This is an incredibly short wavelength, not only relative to the length of the line but absolutely as well. No experiment could detect the physical implications of such a short wavelength. It is indeed many, many times smaller than the diameter of the mass itself. For example, to observe a diffraction effect for such particles the spacings in the grating must be of the order of magnitude of 1×10^{-27} cm. Such a grating cannot be made from ordinary matter since atoms themselves are about 10^{19} times larger than this. Even if such a grating could be found, it certainly wouldn't affect the motion of a mass of 1 g as the size of the mass is approximately 10^{28} times larger than the spacings in the grating! Clearly, quantum effects will not be observed for massive particles. It is also clear that the factor which determines when quantum effects will be observed and when they will be absent is the magnitude of Planck's constant h. The very small magnitude of h restricts the observation of quantum effects to the realm of small masses.

The Hydrogen Atom

The study of the hydrogen atom is more complicated than our previous example of an electron confined to move on a line. Not only does the motion of the electron occur in three dimensions but there is also a force acting on the electron. This force, the electrostatic force of attraction, is responsible for holding the atom together. The magnitude of this force is given by the product of the nuclear and electronic charges divided by the square of the distance between them. In the previous example of an electron confined to move on a line, the total energy was entirely kinetic in origin since there were no forces acting on the electron. In the hydrogen atom however, the energy of the electron, because of the force exerted on it by the nucleus, will consist of a potential energy (one which depends on the position of the electron relative to the nucleus) as well as a kinetic energy. The potential energy arising from the force of attraction between the nucleus and the electron is:

$$PE = e^2/r$$

Let us imagine for the moment that the proton and the electron behave classically. Then, if the nucleus is held fixed at the origin and the electron allowed to move relative to it, the potential energy would vary in the manner indicated in Fig. The potential energy is independent of the direction in space and depends only on the distance r between the electron and the nucleus. Thus Fig. refers to any line directed from the nucleus to the electron. The r-axis in the figure may be taken literally as a line through the nucleus. Whether the electron moves to the right or to the left the potential energy varies in the same manner.

The potential energy is zero when the two particles are very far apart ($r = \infty$), and equals minus infinity when r equals zero. We shall take the energy for $r = \infty$ as our zero of energy. Every energy will be measured relative to this value. When a stable atom is formed, the electron is attracted to the nucleus, r is less than infinity, and the energy will be negative. A negative value for the energy implies that energy must be supplied to the system if the electron is to overcome the attractive force of the nucleus and escape from the atom. The electron has again "fallen into a potential well." However, the shape of the well is no longer a simple square one as previously considered for an electron confined to move on a line, but has the shape shown in Fig. This shape is a consequence of there being a force acting on the electron and hence a potential energy contribution which depends on the distance between the two particles. This is the nature of the problem. Now let us see what quantum mechanics predicts for the motion of the electron in such a situation.

The Quantization of Energy

The motion of the electron is not free. The electron is bound to the atom by the attractive force of the nucleus and consequently quantum mechanics predicts that the total energy of the electron is quantized. The expression for the energy is:

$$E_n = \frac{-2\pi^2 me^2 Z^2}{n^2 h^2} \qquad n = 1, 2, 3, \ldots\ldots$$

where m is the mass of the electron, e is the magnitude of the electronic charge, n is a quantum number, h is Planck's constant and Z is the atomic number (the number of positive charges in the nucleus).

This formula applies to any one-electron atom or ion. For example, He^+ is a one-electron system for which $Z = 2$. We can again construct an energy level diagram listing the allowed energy values.

These are obtained by substituting all possible values of n into equation (1). As in our previous example, we shall represent all the constants which appear in the expression for E_n by a constant K and we shall set $Z = 1$.

$$E_n = -K/n^2 \qquad n = 1, 2, 3, \ldots\ldots$$

Since the motion of the electron occurs in three dimensions we might correctly anticipate three quantum numbers for the hydrogen atom. But the energy depends only on the quantum number n and for this reason it is called the *principal quantum number*. In this case, the energy is inversly dependent upon n^2, and as n is increased the energy becomes less negative with the spacings between the energy levels decreasing in size. When $n = \infty$, $E = 0$ and the electron is free of the attractive force of the nucleus. The average distance between the nucleus and the electron (the average value of r) increases as the energy or the value of n increases. Thus energy must be supplied to pull the electron away from the nucleus.

The parallelism between increasing energy and increasing average value of r is a useful one. In fact, when an electron loses energy, we refer to it as "falling" from one energy level to a lower one on the energy level diagram. Since the average distance between the nucleus and the electron also decreases with a decrease in n, then the electron literally does fall in closer to the nucleus when it "falls" from level to level on the energy level diagram.

The energy difference between E_∞ and E_1:

$$E_\infty - E_1 = 0-(-K) = K = \frac{2\pi^2 me^4}{h^2} = \frac{e^2}{2a_0}$$

is called the ionization energy. It is the energy required to pull the electron completely away from the nucleus and is, therefore, the energy of the reaction:

$$H \rightarrow H^{+} + e^{-} \qquad \Delta E = K = 13.60\,ev = \frac{1}{2}au$$

This amount of energy is sufficient to separate the electron from the attractive influence of the nucleus and leave both particles at rest. If an amount of energy greater than K is supplied to the electron, it will not only escape from the atom but the energy in excess of K will appear as kinetic energy of the electron. Once the electron is free it may have any energy because all velocities are then possible. This is indicated in the energy level diagram by the shading above the $E_{\infty} = 0$ line. An electron which possesses and energy in this region of the diagram is a free electron and has kinetic energy of motion only.

THE HYDROGEN ATOM SPECTRUM

As mentioned earlier, hydrogen gas emits coloured light when a high voltage is applied across a sample of the gas contained in a glass tube fitted with electrodes. The electrical energy transmitted to the gas causes many of the hydrogen molecules to dissociate into atoms:

$$H_2 \rightarrow H + H$$

The electrons in the molecules and in the atoms absorb energy and are excited to high energy levels. Ionization of the gas also occurs. When the electron is in a quantum level other than the lowest level (with $n = 1$) the electron is said to be excited, or to be in an excited level. The lifetime of such an excited level is very brief, being of the order of magnitude of only 10^{-8} sec. The electron loses the energy of excitation by falling to a lower energy level and at the same time emitting a photon to carry off the excess energy. We can easily calculate the frequencies which should appear in the emitted light by calculating the difference in energy between the two levels and making use of Bohr's frequency condition:

$$\nu = \frac{\varepsilon}{h}$$

$$= \frac{E_{n'} - E_n}{h} \qquad n' > n$$

Suppose we consider all those frequencies which appear when the electron falls to the lowest level, $n = 1$,

$$\nu = \frac{\varepsilon}{h}$$

$$= \frac{E_n - E_n}{h}$$

$$= \frac{K}{h}\left(\frac{1}{1} - \frac{1}{n^2}\right) \qquad n = 1, 2, 3, \ldots\ldots$$

Every value of n substituted into this equation gives a distinct value for ν. In Fig. 3-3 we illustrate the changes in energy which result when the electron emits a photon by an arrow connecting the excited level (of energy E_n) with the ground level (of energy E_1). The frequency resulting from each drop in energy will be directly proportional to the length of the arrow. Just as the arrows increase in length as n is increased, so ν increases.

Each line in the spectrum is placed beneath the arrow which represents the change in energy giving rise to that particular line. Free electrons with varying amounts of kinetic energy ($\frac{1}{2}mu^2$) can also fall to the $n = 1$ level.

The energy released in the reversed ionization reaction:

$$H^+ + e^- \rightarrow H$$

will equal K, the difference between E_∞ and E_1, plus $\frac{1}{2}m\upsilon^2$, the kinetic energy originally possessed by the electron. Since this latter energy is not quantized, every energy value greater than K should be possible and every frequency greater than that corresponding to

$$\nu = \frac{K}{h}$$

should be observed. The line spectrum should, therefore, collapse into a continuous spectrum at its high frequency end. Thus the energy continuum above E_∞ gives rise to a continuum of frequencies in the emission spectrum. The beginning of the continuum should be the frequency corresponding to the jump from E_∞ to E_1, and thus we can determine K, the ionization energy of the hydrogen atom, from the observation of this frequency. Indeed, the spectroscopic method is one of the most accurate methods of determining ionization energies.

The hydrogen atom does possess a spectrum identical to that predicted by equation, and the observed value for K agrees with the theoretical value. This particular series of lines, called the Lyman series, falls in the ultraviolet region of the spectrum because of the large energy changes involved in the transitions from the excited levels to the lowest level. The first few members of a second series of lines, a second line spectrum, falls in the visible portion of the spectrum. It is called the Balmer series and arises from electrons in excited levels falling to the second quantum level. Since E_2 equals only one quarter of E_1, the energy jumps are smaller and the frequencies are correspondingly lower than those observed in the Lyman series. Four lines can be readily seen in this series: red, green, blue, and violet. Each colour results from the electrons falling from a specific level, to the $n = 2$ level: red $E_3 \rightarrow E_2$; green, $E_4 \rightarrow E_2$; blue, $E_5 \rightarrow E_2$; and violet $E_6 \rightarrow E_2$. Other series, arising from electrons falling to the $n = 3$ and $n = 4$ levels, can be found in the infrared (frequencies preceding the red end or long wavelength end of the visible spectrum).

The fact that the hydrogen atom exhibits a line spectrum is visible proof of the quantization of energy on the atomic level.

THE PROBABILITY DISTRIBUTIONS FOR THE HYDROGEN ATOM

To what extent will quantum mechanics permit us to pinpoint the position of an electron when it is bound to an atom? We can obtain an order of magnitude answer to this question by applying the uncertainty principle

$$\Delta x \Delta p = h$$

to estimate Δx. The value of Δx will represent the minimum uncertainty in our knowledge of the position of the electron. The momentum of an electron in an atom is of the order of magnitude of 9×10^{-19} g cm/sec. The *uncertainty* in the momentum Δp must necessarily be of the same order of magnitude. Thus

$$\Delta x = \frac{7\times10^{-27}}{9\times10^{-19}} \sim 10^{-8}$$

The uncertainty in the position of the electron is of the same order of magnitude as the diameter of the atom itself. As long as the electron is bound to the atom, we will not be able to say much more about its position than that it is in the atom. Certainly all models of the atom which describe the electron as a particle following a definite trajectory or orbit must be discarded.

We can obtain an energy and one or more wave functions for every value of n, the principal quantum number, by solving Schrödinger's equation for the hydrogen atom. A knowledge of the wave functions, or probability amplitudes ψ_n, allows us to calculate the probability distributions for the electron in any given quantum level. When $n = 1$, the wave function and the derived probability function are independent of direction and depend only on the distance r between the electron and the nucleus. In Fig, we plot both ψ_1 and P_1 versus r, showing the variation in these functions as the electron is moved further and further from the nucleus in any one direction. (These and all succeeding graphs are plotted in terms of the atomic unit of length, $a_0 = 0.529 \times 10^{-8}$ cm.)

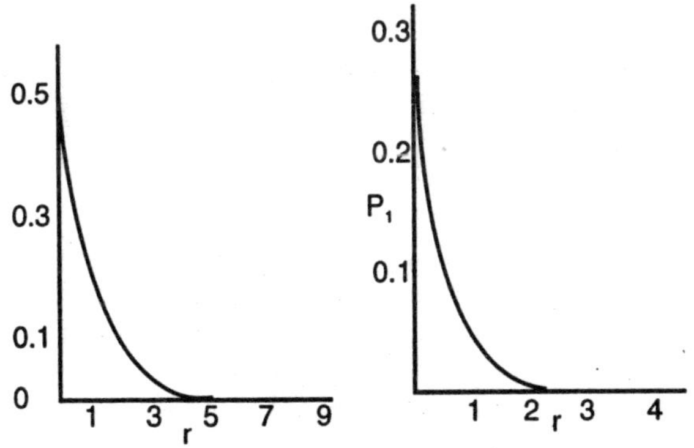

Fig. The wave function and probability distribution as functions of r for the $n = 1$ level of the H atom. The functions and the radius r are in atomic units in this and succeeding figures.

Two interpretations can again be given to the P_1 curve. An experiment designed to detect the position of the electron with an uncertainty much less than the diameter of the atom itself (using light of short wavelength) will, if repeated a large number of times, result in Fig. for P_1. That is, the electron will be detected close to the nucleus most frequently and the probability of observing it at some distance from the nucleus will decrease rapidly with increasing r. The atom will be ionized in making each of these observations because the energy of the photons with a wavelength much less than 10^{-8} cm will be greater than K, the amount of energy required to ionize the hydrogen atom. If light with a wavelength comparable to the diameter of the atom is employed in the experiment, then the electron will not be excited but our knowledge of its position will be correspondingly less precise. In these experiments, in which the electron's energy is not changed, the electron will appear to be "smeared out" and we may interpret P_1 as giving the fraction of the total electronic charge to be found in every small volume element of space.

When the electron is in a definite energy level we shall refer to the P_n distributions as *electron density distributions*, since they describe the manner in which the total electronic charge is distributed in space. The electron density is expressed in terms of the number of electronic charges per unit volume of space, e^-/V. The volume V is usually expressed in atomic units of length cubed, and one atomic unit of electron density is then e^-/a_0^3. To give an idea of the order of magnitude of an atomic density unit, 1 au of charge density $e^-/a_0^3 = 6.7$ electronic charges per cubic Ångstrom. That is, a cube with a length of 0.52917×10^{-8} cm, if uniformly filled with an electronic charge density of 1 au, would contain 6.7 electronic charges.

P_1 may be represented in another manner. Rather than considering the amount of electronic charge in one particular small element of space, we may determine the total amount of charge lying within a thin spherical shell of space. Since the distribution is independent of direction, consider adding up all the charge density which lies within a volume of space bounded by an inner sphere of radius r and an outer concentric sphere with a radius only infinitesimally greater, say $r + \Delta r$. The area of the inner sphere is

$4\pi r^2$ and the thickness of the shell is Δr. Thus the volume of the shell is $4\pi r^2 \Delta r$ and the product of this volume and the charge density $P_1(r)$, which is the charge or number of electrons per unit volume, is therefore the total amount of electronic charge lying between the spheres of radius r and $r + \Delta r$. The product $4\pi r^2 P_n$ is given a special name, the radial distribution function, which we shall label $Q_n(r)$.

The radial distribution function is plotted in Fig for the ground state of the hydrogen atom.

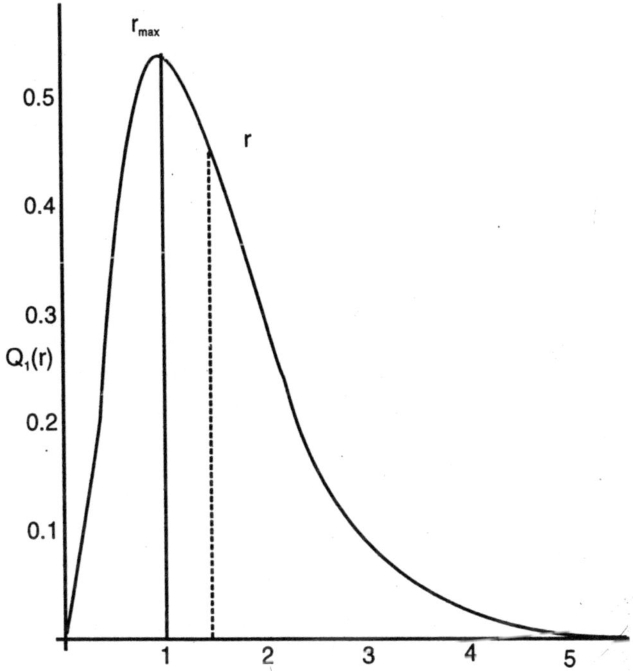

Fig. The radial distribution function $Q_1(r)$ for an H atom. The value of this function at some value of r when multiplied by Dr gives the number of electronic charges within the thin shell of space lying between spheres of radius r and r + Dr.

The curve passes through zero at $r = 0$ since the surface area of a sphere of zero radius is zero. As the radius of the sphere is increased, the volume of space defined by $4\pi r^2 \Delta r$ increases. However, as shown in Fig, the absolute value of the electron density at a given point decreases with r and the resulting curve must pass through a maximum. This maximum occurs at $r_{max} = a_0$. Thus more of the electronic charge is present at a distance a_0, out from the nucleus than at any other value of r. Since the curve is unsymmetrical, the average value of r, denoted by $\bar{r}$, is not equal to r_{max}. The average value of r is indicated on the figure by a dashed line. A "picture" of the electron density distribution for the electron in the $n = 1$ level of the hydrogen atom would be a spherical ball of charge, dense around the nucleus and becoming increasingly diffuse as the value of r is increased.

We could also represent the distribution of negative charge in the hydrogen atom in the manner used previously for the electron confined to move on a plane, Fig., by displaying the charge density in a plane by means of a contour map. Imagine a plane through the atom including the nucleus. The density is calculated at every point in this plane. All points having the same value for the electron density in this plane are joined by a contour line. Since the electron density depends only on r, the distance from the nucleus, and not on the direction in space, the contours will be circular. A contour map is useful as it indicates the "shape" of the density distribution

This completes the description of the most stable state of the hydrogen atom, the state for which $n = 1$. Before proceeding with a discussion of the excited states of the hydrogen atom we must introduce a new term. When the energy of the electron is increased to another of the allowed values, corresponding to a new value for n, ψ_n and P_n change as well. The wave functions ψ_n for the hydrogen atom are given a special name, *atomic orbitals*, because they play such an important role in all of our future discussions of the electronic structure of atoms. In general the word orbital is the name given to a wave function which determines the motion of a single electron. If the one-electron wave function is for an atomic system, it is called an atomic orbital.

For every value of the energy E_n, for the hydrogen atom, there is a degeneracy equal to n^2. Therefore, for $n = 1$, there is but one atomic orbital and one electron density distribution. However, for $n = 2$, there are four different atomic orbitals and four different electron density distributions, all of which possess the same value for the energy, E_2. Thus for all values of the principal quantum number n there are n^2 different ways in which the electronic charge may be distributed in three-dimensional space and still possess the same value for the energy. For every value of the principal quantum number, *one* of the possible atomic orbitals is independent of direction and gives a spherical electron density distribution which can be represented by circular contours as has been exemplified above for the case of $n = 1$. The other atomic orbitals for a given value of n exhibit a directional dependence and predict density distributions which are not spherical but are concentrated in planes or along certain axes. The angular dependence of the atomic orbitals for the hydrogen atom and the shapes of the contours of the corresponding electron density distributions are intimately connected with the angular momentum possessed by the electron.

The physical quantity known as angular momentum plays a dominant role in the understanding of the electronic structure of atoms. To gain a physical picture and feeling for the angular momentum it is necessary to consider a model system from the classical point of view. The simplest classical model of the hydrogen atom is one in which the electron moves in a circular orbit with a constant speed or angular velocity. Just as the ordinary momentum *m*v plays a dominant role in the analysis of straight line or linear motion, so angular momentum plays the central role in the analysis of a system with circular motion as found in the model of the hydrogen atom.

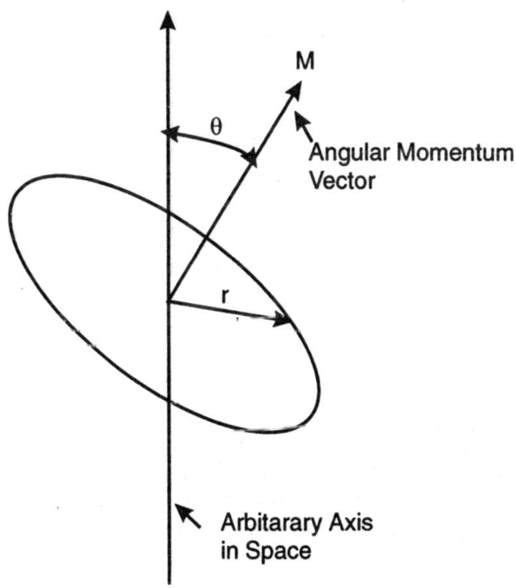

Fig. The Angular Momentum Vector for a Classical Model of The Atom.

In Fig, *m* is the mass of the electron, *v* is the linear velocity (the velocity the electron would possess if it continued moving at a tangent to the orbit as indicated in the figure) and *r* is the radius of the orbit. The linear velocity v is a vector since it possesses at any instant both a magnitude and a direction in space. Obviously, as the electron rotates in the orbit the direction of *v* is constantly changing, and thus the linear momentum *mv* is *not* constant for the circular motion. This is so even though the speed of the electron (the magnitude of *v* which is denoted by υ) remains unchanged. According to Newton's second law, a force must be acting on the electron if its momentum changes with time. This is the force which prevents the electron from flying on tangent to its orbit. In an atom the attractive force which contains the electron is the electrostatic force of attraction between the nucleus and the electron, directed along the radius *r* at right angles to the direction of the electron's motion.

The angular momentum, like the linear momentum, is a vector and is defined as follows:

magnitude of angular momentum = $M = m\upsilon r$

The angular momentum vector *M* is directed along the axis of rotation. From the definition it is evident that the angular momentum vector will remain constant as long as the speed of the electron in the orbit is constant (υ remains unchanged) and the plane and radius of the orbit remain unchanged. Thus for a given orbit, the angular momentum is constant as long as the angular velocity of the particle in the orbit is constant. In an atom the only force on the electron in the orbit is directed along *r*; it has no component in the direction of the motion. The force acts in such a way as to change only the linear momentum. Therefore, while the linear momentum is not constant during the circular motion, the angular momentum is. A force exerted on the particle in the direction of the vector *v* would change the angular velocity and the angular momentum. When a force is applied which does change *M*, a *torque* is said to be acting on the system. Thus angular momentum and torque are related in the same way as are linear momentum and force.

The important point of the above discussion is that both the angular momentum and the energy of an atom remain constant if the atom is left undisturbed. *Any physical quantity which is constant in a classical system is both conserved and quantized in a quantum mechanical system.* Thus both the energy and the angular momentum are quantized for an atom.

There is a quantum number, denoted by l, which governs the magnitude of the angular momentum, just as the quantum number n determines the energy. The *magnitude* of the angular momentum may assume only those values given by:

$$M = \sqrt{l(l+1)}\,(h/2\pi), \text{where } l = 1, 2, 3, \ldots\ldots n-1$$

Furthermore, the value of n limits the maximum value of the angular momentum as the value of l cannot be greater than $n-1$. For the state $n = 1$ discussed above, l may have the value of zero only. When $n = 2$, l may equal 0 or 1, and for $n = 3$, $l = 0$ or 1 or 2, etc. When $l = 0$, it is evident from equation that the angular momentum of the electron is zero. The atomic orbitals which describe these states of zero angular momentum are called *s* orbitals. The *s* orbitals are distinguished from one another by stating the value of n, the principal quantum number. They are referred to as the 1*s*, 2*s*, 3*s*, etc., atomic orbitals.

The preceding discussion referred to the 1*s* orbital since for the ground state of the hydrogen atom $n = 1$ and $l = 0$. This orbital, and all *s* orbitals in general, predict spherical density distributions for the electron as exemplified by Fig. for the 1*s* density. Figure shows the radial distribution functions $Q(r)$ which apply when the electron is in a 2*s* or 3*s* orbital to illustrate how the character of the density distributions change as the value of n is increased.

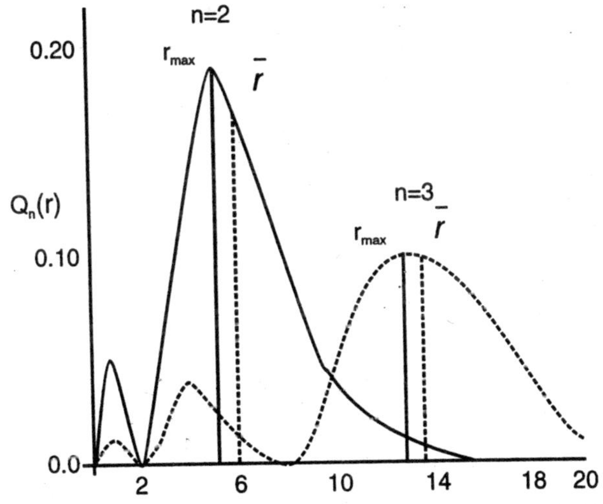

Fig. Radial Distribution Functions for the 2*s* and 3*s* Density Distributions.

Comparing these results with those for the 1*s* orbital in Fig we see that as *n* increases the average value of *r* increases. This agrees with the fact that the energy of the electron also increases as *n* increases. The increased energy results in the electron being on the average pulled further away from the attractive force of the nucleus. As in the simple example of an electron moving on a line, nodes (values of *r* for which the electron density is zero) appear in the probability distributions. The number of nodes increases with increasing energy and equals $n - 1$.

When the electron possesses angular momentum the density distributions are no longer spherical. In fact for each value of *l*, the electron density distribution assumes a characteristic shape.

When $l = 1$, the orbitals are called *p* orbitals. In this case the orbital and its electron density are concentrated along a line (axis) in space. The 2*p* orbital or wave function is positive in value on one side and negative in value on the other side of a plane which is perpendicular to the axis of the orbital and passes through the nucleus. The orbital has a node in this plane, and consequently an electron in a 2*p* orbital does not place any electronic charge density at the nucleus. The electron density of a 1*s* orbital, on the other hand, is a maximum at the nucleus. The same diagram for the 2*p* density distribution is obtained for any plane which contains this axis. Thus in three dimensions the electron density would appear to be concentrated in two lobes, one on each side of the nucleus, each lobe being circular in cross section.

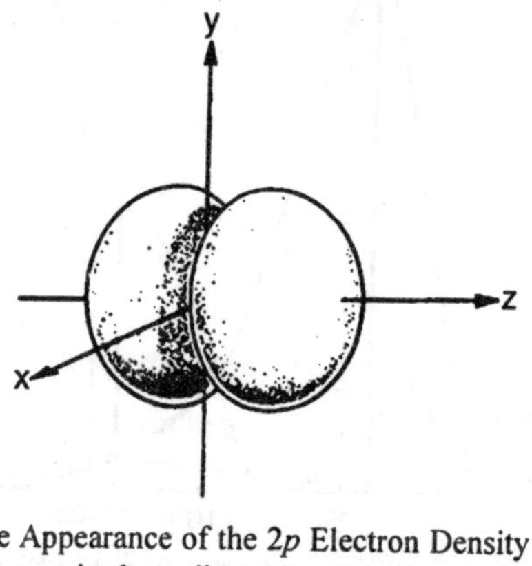

Fig. The Appearance of the 2*p* Electron Density Distribution in three-dimensional Space.

When $l = 2$, the orbitals are called *d* orbitals and Fig. shows the contours in a plane for a 3*d* orbital and its density distribution. Notice that the density is again zero at the nucleus and that there are now two nodes in the orbital and in its density distribution. As a final example, Fig shows the contours of the orbital and electron density distribution obtained for a 4*f* atomic orbital which occurs when $n = 4$ and $l = 3$. The point to notice is that as the angular momentum of the electron increases, the density distribution becomes increasingly concentrated along an axis or in a plane in space. Only electrons in *s* orbitals with zero angular momentum give spherical density distributions and in addition place charge density at the position of the nucleus.

We have not as yet accounted for the full degeneracy of the hydrogen atom orbitals which we stated earlier to be n^2 for every value of *n*. For example, when $n = 2$, there are four distinct atomic orbitals. The remaining degeneracy is again determined by the angular momentum of the system. Since angular momentum like linear momentum is a vector quantity, we may refer to the component of the angular momentum vector which lies along some chosen axis. For reasons we shall investigate, the number of values a particular component can assume for a given value of *l* is $(2l + 1)$. Thus when $l = 0$, there is no angular momentum and there is but a single orbital, an *s* orbital. When $l = 1$, there are three possible values for the component $(2 \times 1 + 1)$ of the total angular momentum which are physically distinguishable from one another. There are, therefore, three *p* orbitals. Similarly there are five *d* orbitals, $(2 \times 2+1)$, seven *f* orbitals, $(2 \times 3 +1)$, etc. All of the orbitals with the same value of *n* and *l*, the three 2*p* orbitals for example, are similar but differ in their spatial orientations.

To gain a better understanding of this final element of degeneracy, we must consider in more detail what quantum mechanics predicts concerning the angular momentum of an electron in an atom.

ANGULAR MOMENTUM OF AN ELECTRON IN AN H ATOM

The simplest classical model of the hydrogen atom is one in which the electron moves in a circular planar orbit about the nucleus as previously discussed and as illustrated in Fig. The angular momentum vector M in this figure is shown at an angle θ with respect to some arbitrary axis in space. Assuming for the moment that we can somehow physically define such an axis, then in the classical model of the atom there should be an infinite number of values possible for the component of the angular momentum vector along this axis. As the angle between the axis and the vector M varies continuously from 0°, through 90° to 180°, the component of M along the axis would vary correspondingly from M to zero to $-M$. Thus the quantum mechanical statements regarding the angular momentum of an electron in an atom differ from the classical predictions in two startling ways. First, the magnitude of the angular momentum (the length of the vector M) is restricted to only certain values given by:

$$\sqrt{1(1+1)}\frac{h}{2\pi}, \text{ where } l = 1, 2, 3, \ldots\ldots$$

The magnitude of the angular momentum is quantized. Secondly, quantum mechanics states that the component of M along a given axis can assume only $(2l + 1)$ values, rather than the infinite number allowed in the classical model. In terms of the classical model this would imply that when the magnitude of M is $\sqrt{2}(h/2\pi)$ (the value when $l = 1$), there are only three allowed values for θ, the angle of inclination of M with respect to a chosen axis.

The angle θ is another example of a physical quantity which in a classical system may assume any value, but which in a quantum system may take on only certain discrete values. You need not accept this result on faith. There is a simple, elegant experiment which illustrates the "quantization" of θ, just as a line spectrum illustrates the quantization of the energy.

If we wish to measure the number of possible values which the component of the angular momentum may exhibit with respect to some axis we must first find some way in which we can physically define a direction or axis in space. To do this we make use of the magnetism exhibited by an electron in an atom. The flow of electrons through a loop of wire (an electric current) produces a magnetic field. At a distance from the ring of wire, large compared to the diameter of the ring, the magnetic field produced by the current appears to be the same as that obtained from a small bar magnet with a north pole and a south pole. Such a small magnet is called a magnetic dipole, i.e., two poles separated by a small distance.

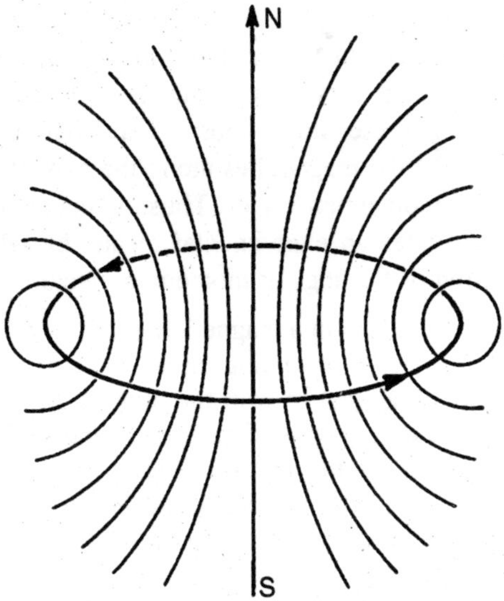

Fig. The Magnetic field Produced by a Current in a Loop of Wire

The electron is charged and the motion of the electron in an atom could be thought of as generating a small electric current. Associated with this current there should be a small magnetic field. The magnitude of this magnetic field is related to the angular momentum of the electron's motion in roughly the same way that the magnetic field produced by a current in a loop of wire is proportional to the strength of the current flowing in the wire.

The strength of the atomic magnetic dipole is given by μ where:

(5) $$\mu = \sqrt{1+(l+1)}\beta_m$$

Just as there is a fundamental unit of negative charge denoted by e^- so there is a fundamental unit of magnetism at the atomic level denoted by β_m and called the Bohr magneton. From equation we can see that the strength of the magnetic dipole will increase as the angular momentum of the electron increases. This is analogous to increasing the magnetic field by increasing the strength of the current through a circular loop of wire The magnetic dipole, since it has a north and a south pole, will define some direction in space. The axis of the magnetic dipole in fact coincides with the direction of the angular momentum vector. Experimentally, a collection of atoms behave as though they were a collection of small bar magnets if the electrons in these atoms possess angular momentum. In addition, the axis of the magnet lies along the axis of rotation, i.e., along the angular momentum vector. Thus the magnetism exhibited by the atoms provides an experimental means by which we may study the direction of the angular momentum vector.

If we place the atoms in a magnetic field they will be attracted or repelled by this field, depending on whether or not the atomic magnets are aligned against or with the applied field. The applied magnetic field *will determine a direction in space*. By measuring the deflection of the atoms in this field we can determine the directions of their magnetic moments and hence of their angular momentum vectors with respect to this applied field. Consider an evacuated tube with a tiny opening at one end through which a stream of atoms may enter. By placing a second small hole in front of the first, inside the tube, we will obtain a narrow beam of atoms which will pass the length of the tube and strike the opposite end. If the atoms possess magnetic moments the path of the beam can be deflected by placing a magnetic field across the tube, perpendicular to the path of the atoms. The magnetic field must be one in which the lines of force diverge thereby exerting an unbalanced force on any magnetic material lying inside the field. This inhomogeneous magnetic field could be obtained through the use of N and S poles of the kind illustrated in Fig.. The direction of the magnetic field will be taken as the direction of the *z*-axis.

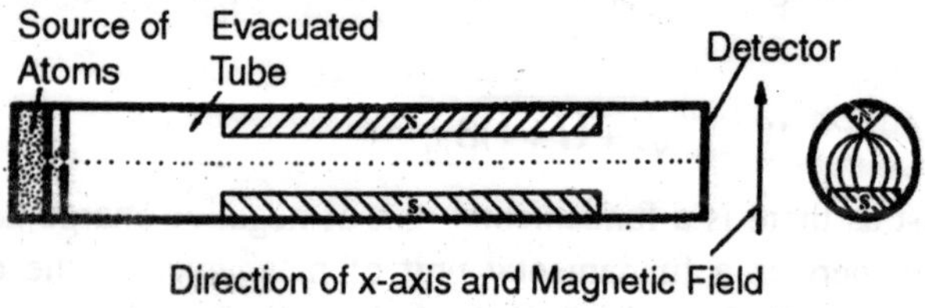

Fig. The Atomic Beam Apparatus

Let us suppose the beam consists of neutral atoms which possess $\sqrt{2}(h/2\pi)$ units of electronic angular momentum (the angular momentum quantum number $l = 1$). When no magnetic field is present, the beam of atoms strikes the end wall at a single *point* in the middle of the detector. What happens when the magnetic field is present? We must assume that before the beam enters the magnetic field, the axes of the atomic magnets are randomly oriented with respect to the z-axis. According to the concepts of classical mechanics, the beam should spread out along the direction of the magnetic field and produce a line rather than a point at the end of the tube. Actually, the beam is split into three distinct component beams each of equal intensity producing three spots at the end of the tube.

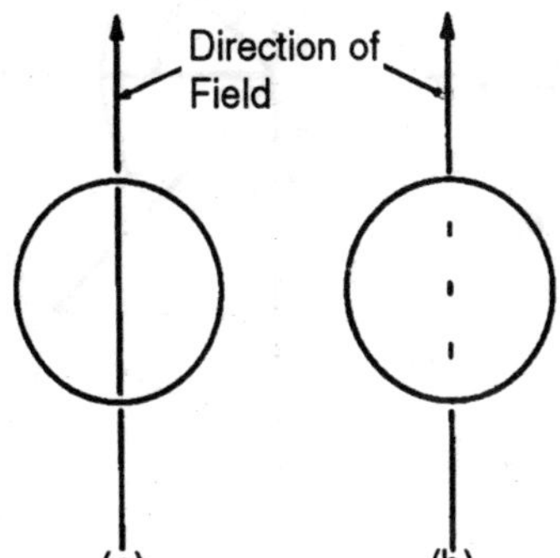

Fig. (a) The Result of The Atomic Beam Experiment as Predicted by Classical Mechanics,

(b) The Observed Result of the Atomic Beam Experiment

The startling results of this experiment can be explained only if we assume that while in the magnetic field each atomic magnet could assume only one of three possible orientations with respect to the applied magnetic field.

The atomic magnets which are aligned perpendicular to the direction of the field are not deflected and will follow a straight path through the tube. The atoms which are attracted upwards must have their magnetic moments oriented. From the known strength of the applied inhomogeneous magnetic field and the measured distance through which the beam has been deflected upwards, we can determine that the component of the magnetic moment lying along the *z*-axis is only β_m in magnitude rather than the value of $\sqrt{2}\beta_m$ This latter value would result if the axis of the atomic magnet was parallel to the *z*-axis, i.e., the angle θ = 0°.

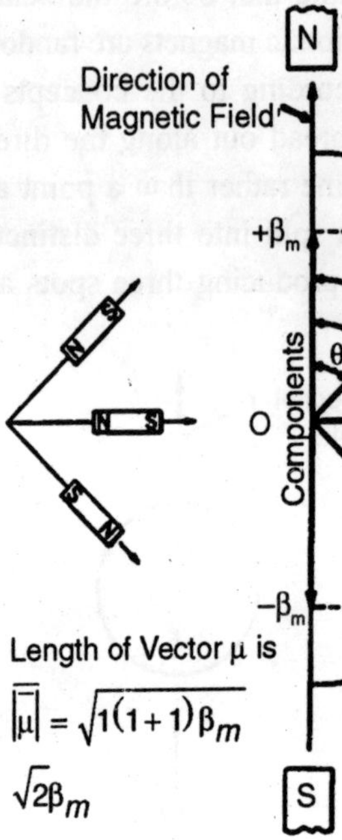

Fig. The three possible orientations for the total magnetic moment with respect to an external magnetic field for an atom with $l = 1$

Instead θ assumes a value such that the component of the total moment $\left(\sqrt{l(l+1)}\beta_m\right)$ lying along the *z*-axis is just $l\beta_m$. Similarly the beam which is deflected downwards possesses a magnetic moment along the *z*-axis of $-\beta_m$ or $-l\beta_m$. The classical prediction for this experiment assumes that θ may equal all values from 0° to 180°, and thus all values (from a maximum of $-\sqrt{2}\beta_m$ (θ = 0°) to 0 (θ =90°) to $-\sqrt{2}\beta_m$ (θ = 180°)) for the component of the atomic moment along the *z*-axis would be possible. Instead, θ is found to equal only those values such that the magnetic moment along the *z*-axis equals $+\beta_m$, 0 and $-\beta_m$.

The angular momentum of the electron determines the magnitude and the direction of the magnetic dipole. (Recall that the vectors for both these quantities lie along the same axis.) Thus the number of possible values which the component of the angular momentum vector may assume along a given axis must equal the number of values observed for the component of the magnetic dipole along the same axis. In the present example the values of the angular momentum component are $+1(h/2\pi)$, 0 and $-1(h/2\pi)$, or since $l = 1$ in this case, $+\ l(h/2\pi)$, 0 and $-l(h/2\pi)$. In general, it is found that the number of observed values is always $(2l + 1)$ the values being:

$$l\left(\frac{h}{2\pi}\right),(l-1)\left(\frac{h}{2\pi}\right),...0...,-1\left(\frac{h}{2\pi}\right)$$

for the angular momentum and:

$$l\beta_m,(l-1)\beta_m,..0,...,-l\beta_m$$

for the magnetic dipole. The number governing the magnitude of the component of M and $\bar{\mu}$, ranges from a maximum value of l and decreases in steps of unity to a minimum value of $-l$. This number is the third and final quantum number which determines the motion of an electron in a hydrogen atom. It is given the symbol m and is called the magnetic quantum number.

In summary, the angular momentum of an electron in the hydrogen atom is quantized and may assume only those values given by:

$$\sqrt{1(1+1)}\frac{h}{2\pi}, \text{ where } l = 1, 2, 3, \text{.....}, n-1$$

Furthermore, it is an experimental fact that the component of the angular momentum vector along a given axis is limited to $(2l + 1)$ different values, and that the magnitude of this component is quantized and governed by the quantum number m which may assume the values $l, l-1, \ldots, 0, \ldots, -l$. These facts are illustrated in Fig for an electron in a d orbital in which $l = 2$.

$$|M| = \sqrt{2(2+1)} = \sqrt{6}$$

The quantum number m determines the magnitude of the component of the angular momentum along a given axis in space. Therefore, it is not surprising that this same quantum number determines the axis along which the electron density is concentrated. When $m = 0$ for a p electron (regardless of the n value, $2p$, $3p$, $4p$, etc.) the electron density distribution is concentrated along the z-axis implying that the classical axis of rotation must lie in the x-y plane. Thus a p electron with $m = 0$ is most likely to be found along one axis and has a zero probability of being on the remaining two axes. The effect of the angular momentum possessed by the electron is to concentrate density along one axis. When $m = 1$ or -1 the density distribution of a p electron is concentrated in the x-y plane with doughnut-shaped circular contours. The $m = 1$ and -1 density distributions are identical in appearance. Classically they differ only in the direction of rotation of the electron around the z-axis; counter-clockwise for $m = +1$ and clockwise for $m = -1$. This explains why they have magnetic moments with their north poles in opposite directions.

We can obtain density diagrams for the $m = +1$ and -1 cases similar to the $m = 0$ case by removing the resultant angular momentum component along the z-axis. We can take combinations of the $m = +1$ and -1 functions such that one combination is concentrated along the x-axis and the other along the y-axis, and both are identical to the $m = 0$ function in their appearance. Thus these functions are often labelled as p_x, p_y and p_z functions rather than by their m values. The m value is, however, the true quantum number and we are cheating physically by labelling them p_x, p_y and p_z. This would correspond to applying the field first in the z direction, then in the x direction and finally in the y direction and trying to save up the information each time. In reality when the direction of the field is changed, all the information regarding the previous direction is lost and every atom will again align itself with one chance out of three of being in one of the possible component states with respect to the new direction.

We should note that the r dependence of the orbitals changes with changes in n or l, but the directional component changes with *l* and *m* only. Thus all *s* orbitals possess spherical charge distributions and all *p* orbitals possess dumb-bell shaped charge distributions regardless of the value of *n*.

Table. The Atomic Orbitals for the Hydrogen Atom

E_n	n	l	m	Symbol for orbital	
–K	1	0	0	1s	
	2	0	0	2s	
	2	1	1	$2p_{+1}$	
	2	1	0	$2p_0$	p_x, p_y, p_z
	2	1	–1	$2p_{-1}$	
	3	0	0	3s	
	3	1	1	$3p_{+1}$	
	3	1	0	$3p_0$	p_x, p_y, p_z
	3	1	–1	$3p_{-1}$	
	3	2	2	$3d_{+2}$	
	3	2	1	$3d_{+1}$	
	3	2	0	$3d_0$	
	3	2	–1	$3d_{-1}$	
	3	2	–2	$3d_{-2}$	

Summarizes the allowed combinations of quantum numbers for an electron in a hydrogen atom for the first few values of *n*; the corresponding name (symbol) is given for each orbital. Notice that there are n^2 orbitals for each value of *n*, all of which belong to the same quantum level and have the same energy. There are *n*–1 values of *l* for each value of *n* and there are (2*l* + 1) values of *m* for each value of *l*. Notice also that for every increase in the value of *n*, orbitals of the same *l* value (same directional dependence) as found for the preceding value of *n* are repeated. In addition, a new value of *l* and a new shape are introduced. Thus there is a repetition in the shapes of the density distributions along with an increase in their number. We can see evidence of a periodicity in these functions (a periodic re-occurrence of a given density distribution) which we might hope to relate to the periodicity observed in the chemical and physical properties of the elements. We might store this idea in the back of our minds until later.

We can summarize what we have found so far regarding the energy and distribution of an electron in a hydrogen atom thus:

1. The energy increases as *n* increases, and depends only on *n*, the principal quantum number.
2. The average value of the distance between the electron and the nucleus increases as *n* increases.
3. The number of nodes in the probability distribution increases as *n* increases.
4. The electron density becomes concentrated along certain lines (or in planes) as *l* is increased.

Some words of caution about energies and angular momentum should be added. In passing from the domain of classical mechanics to that of quantum mechanics we retain as many of the familiar words as possible. Examples are kinetic and potential energies, momentum, and angular momentum. We must, however, be on guard when we use these familiar concepts in the atomic domain. All have an altered meaning. Let us make this clear by considering these concepts for the hydrogen atom.

Perhaps the most surprising point about the quantum mechanical expression for the energy is that it does not involve *r*, the distance between the nucleus and the electron. If the system were a classical one, then we would expect to be able to write the total energy E_n as:

$$E_n = KE$$

$$PE = \frac{1}{2}m\upsilon^2$$

Both the *KE* and *PE* would be functions of *r*, i.e., both would change in value as *r* was changed (corresponding to the motion of the electron). Furthermore, the sum of the *PE* and *KE* must always yield the same value of E_n which is to remain constant.

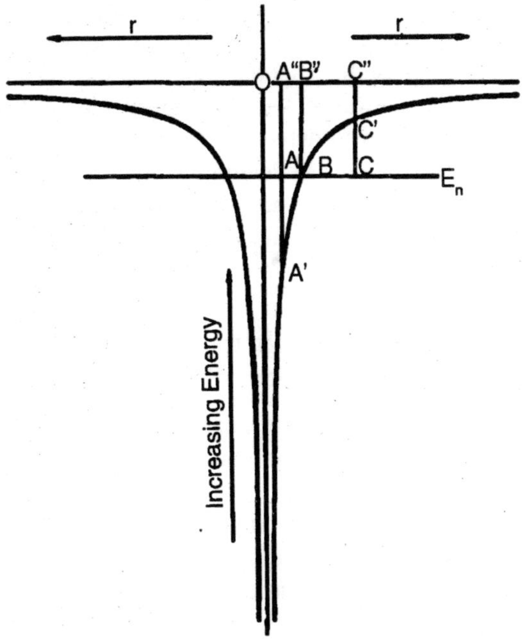

Fig. The potential energy diagram for an *H* atom with one of the allowed energy values superimposed on it.

Fig is the potential energy diagram for the hydrogen atom and we have superimposed on it one of the possible energy levels for the atom, E_n. Consider a classical value for *r* at the point *A*". Classically, when the electron is at the point *A*", its *PE* is given by the value of the *PE* curve at *A*'. The *KE* is thus equal to the length of the line *A*–*A*' in energy units. Thus the sum of *PE* + *KE* adds up to E_n.

When the electron is at the point *B*", its *PE* would equal E_n and its *KE* would be zero. The electron would be motionless. Classically, for this value of E_n the electron could not increase its value of *r* beyond the point represented by *B*". If it did, it would be inside the "potential wall." For example, consider the point *C*". At this value of *r*, the *PE* is given by the value at *C*' which is *now greater* than E_n and hence the *KE* must be equal to the length of the line *C*–*C*'. But the *KE* must now be *negative* in sign so that the sum of *PE* and *KE* will still add up to E_n. What does a negative *KE* mean? It doesn't mean anything as it never occurs in a classical system. Nor does it occur in a quantum mechanical system. It is true that quantum mechanics does predict a finite probability for the electron being inside the potential curve and indeed for all values of r out to infinity. However, the quantum mechanical expression for E_n does not allow us to determine the instantaneous values for the PE and KE. Instead, we can determine only their average values. Thus quantum mechanics does not give equation but instead states only that the *average* potential and kinetic energies may be known:

$$E_n = \overline{PE} = \overline{KE}$$

A bar denotes the fact that the energy quantity has been averaged over the complete motion (all values of r) of the electron.

Why can r not appear in the quantum mechanical expression for E_n, and why can we obtain only average values for the KE and PE? When the electron is in a given energy level its energy is precisely known; it is E_n. The uncertainty in the value of the momentum of the electron is thus at a minimum. Under these conditions we have seen that our knowledge of the position of the electron is very uncertain and for an electron in a given energy level we can say no more about its position than that it is bound to the atom. Thus if the energy is to remain fixed and known with certainty, we cannot, because of the uncertainty principle, refer to (or measure) the electron as being at some particular distance r from the nucleus with some instantaneous values for its PE and KE. Instead, we may have knowledge of these quantities only when they are averaged over all possible positions of the electron. This discussion again illustrates the pitfalls arise when a classical picture of an electron as a particle with a definite instantaneous position is taken literally.

It is important to point out that the classical expressions which we write for the dependence of the potential energy on distance, $-e^2/r$ for the hydrogen atom for example, are the expressions employed in the quantum mechanical calculation. However, only the average value of the PE may be calculated and this is done by calculating the value of $-e^2/r$ at every point in space, taking into account the fraction of the total electronic charge at each point in space. The amount of charge at a given point in three-dimensional space is, of course, determined by the electron density distribution.

Thus the value of $\overline{PE}$ for the ground state of the hydrogen atom is the electrostatic energy of interaction between a nucleus of charge $+1e$ with the surrounding spherical distribution of negative charge.

We can say more about the $\overline{PE}$ and $\overline{KE}$ for an electron in an atom. Not only are these values constant for a given value of *n*, but also for any value of *n*,

$$\overline{KE} = \frac{1}{2}\overline{PE} = -E_n$$

Thus the $\overline{KE}$ is always positive and equal to minus one half of the $\overline{PE}$. Since the total energy E_n is negative when the electron is bound to the atom, we can interpret the stability of atoms as being due to the decrease in the $\overline{PE}$ when the electron is attracted by the nucleus.

The question now arises as to why the electron doesn't "fall all the way" and sit right on the nucleus. When $r = 0$, the $\overline{PE}$ would be equal to minus infinity, and the $\overline{KE}$, which is positive and thus destabilizing, would be zero. Classically this would certainly be the situation of lowest energy and thus the most stable one. The reason for the electron not collapsing onto the nucleus is a quantum mechanical one. If the electron was bound directly to the nucleus with no kinetic energy, its position and momentum would be known with certainty. This would violate Heisenberg's uncertainty principle. The uncertainty principle always operates through the kinetic energy causing it to become large and positive as the electron is confined to a smaller region of space. The smaller the region to which the electron is confined, the smaller is the uncertainty in its position. There must be a corresponding increase in the uncertainty of its momentum. This is brought about by the increase in the kinetic energy which increases the magnitude of the momentum and thus the uncertainty in its value. In other words the bound electron must always possess kinetic energy as a consequence of quantum mechanics.

The $\overline{KE}$ and $\overline{PE}$ have opposite dependences on $\bar{r}$. The $\overline{PE}$ decreases (becomes more negative) as $\bar{r}$ decreases but the increases (making the atom less stable) as $\bar{r}$ decreases. A compromise is reached to make the energy as negative as possible (the atom as stable as possible) and the compromise always occurs when $\overline{KE} = -\frac{1}{2}\overline{PE}$. A further decrease in $\bar{r}$ would decrease the $\overline{PE}$ but only at the expense of a *larger increase* in the $\overline{KE}$. The reverse is true for an increase in $\bar{r}$. Thus the reason the electron doesn't fall onto the nucleus may be summed up by stating that "the electron obeys quantum mechanics, and not classical mechanics."

Some Useful Expressions

Listed below are a number of equations which give the dependence of $\bar{r}$, $\overline{PE}$ and $\overline{KE}$ on the quantum numbers *n*, *l* and *m*. They refer not only to the hydrogen atom but also to any one-electron ion in general with a nuclear charge of *Z*. Thus He^+ is a one-electron ion with $Z = 2$, Li^{+2} another example with $Z = 3$.

The average distance between the electron and the nucleus expressed in atomic units of length is:

$$r_{n,l,m} = \frac{n^2}{Z}\left[1+\frac{1}{2}\left\{1-\frac{l(l-1)}{n^2}\right\}\right]$$

Note that $\bar{r}_{n,l,m}$ is proportional to n^2 for $l = 0$ orbitals, and deviates only slightly from this for $l \neq 0$. The value of $\bar{r}_{n,l,m}$ decreases as *Z* increases because the nuclear attractive force is greater. Thus $\bar{r}_{1,0,0}$ for He^+ would be only one half as large as $\bar{r}_{1,0,0}$ for *H*.

$$E_n = -\frac{Z^2}{n^2}K = -\frac{Z^2}{2n^2}au$$

$$K = \frac{2\pi^2 me^2}{h^2} = \frac{e^2}{2a_0}$$

$$(\overline{PE})_{n,l,m} = -\frac{2z^2}{n^2}K$$

$$(\overline{KE})_{n,l,m} = E_n = +\frac{Z^2}{n^2}K$$

Many-Electron Atoms

The helium atom is a good example of a many-electron atom (that is, an atom which contains more than one electron). No fundamentally new problems are encountered whether we consider two or ten electrons, but a very important problem arises in passing from the one-electron to the two-electron case. To see what this problem is, consider all the potential interactions found in a helium atom. Again, consider the electrons to be point charges and "freeze" them at some instantaneous positions in space.

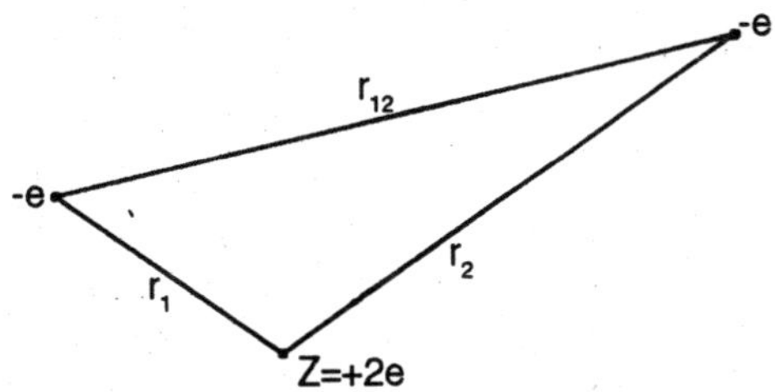

Fig. The potential interactions in an He atom. The electrons are labelled by their charge $-e$, and the nucleus by its charge $Z = +2e$.

The potential energy, the average value of which is to be determined by quantum mechanics, is

$$PE = \frac{(+2e)(-e)}{r_1} + \frac{(+2e)(-e)}{r_2} + \frac{(-e)(-e)}{r_{12}}$$

$$= \frac{-2e^2}{r_1} - \frac{2e^2}{r_1} + \frac{e^2}{r_{12}}$$

The first and second terms in equation represent the attraction of the helium nucleus (for which $Z = 2$) for electrons 1 and 2 respectively. The last term represents the repulsion between the two electrons. It is this last term which makes the problem of the helium atom, and of all many-electron atoms, difficult to solve. No direct solution to the problem exists, the reason being that there are too many interactions to consider simultaneously.

Chapter 3

Atomic Orbital

Since the nuclear charge is twice the electronic charge, the electrostatic energy of repulsion between the electrons will be the smallest of the three terms in equation when the interactions are averaged over all possible positions of the electrons. We could obtain an approximation to the electronic energy of the He atom by neglecting this small term. This is a good idea for another reason. If we ignore the repulsion between the electrons, then physically we are supposing that neither electron "realises" the other electron is present. The problem is thus identical to that of the hydrogen atom, a problem which can be solved exactly, except that the nuclear charge is now +2 rather than +1.

The energy of each electron in the field of a nucleus of charge +2 is determined separately and the total energy is then simply the sum of the two energies. This is the first approximation to the energy. We will also obtain an approximation to the manner in which the electrons are distributed in space. With this latter knowledge, we can estimate the energy of repulsion between the electrons. That is, with the knowledge of how the electrons are distributed relative to one another we can pick up the term in the potential energy we originally neglected, the term e^2/r_{12}, and calculate its contribution to the energy.

This method of calculating the electronic energy of atoms reduces a *many-electron* problem to many *one-electron* problems. Each electron (moving in the attractive field of the nucleus) is treated independently of the others. In addition, since the problem is now a set of one-electron problems, we may carry over and use all of the results obtained for the hydrogen atom.

We pointed out in our discussion of the hydrogen atom that the results we obtained could be applied to any one-electron system by setting Z equal to the appropriate value in all of the formulae. The one-electron energies are easily calculated and are given by $E_n = -\frac{Z^2K}{n^2}$. More important, the concept of atomic orbitals, the one-electron wave functions for the hydrogen atom, may be employed in the many-electron case. When each electron is considered in turn, its motion and distribution in space will again be determined by an atomic orbital. The atomic orbitals will differ from the case of the hydrogen atom in that they will generally be more contracted. The average value of the distance between the nucleus and the electron, , decreased as the nuclear charge and hence the attractive force exerted by the nucleus was increased However, the orbitals will still be determined by the three quantum numbers n, l and m. Increasing Z contracts the orbital, but the symmetry of the problem is left unchanged, i.e., the attraction of the electron by the nucleus is still determined only by the distance between them and does not depend on the direction.

The l and m dependences of the orbitals, which determine the directional properties of the orbital and of the electron density distribution, remain unchanged. Thus we may still refer to "hydrogen-like" $1s$, 2s, $2p$ orbitals For example, the $1s$ orbital is the most stable orbital (most negative E value) for any value of Z, and we naturally assume that the most stable form of the helium atom will be obtained when *both* electrons are placed in the $1s$ orbital. This information, telling us in which atomic orbital the electrons have been placed, is called an *electron configuration.*

An abbreviated notation is used to denote the electron configuration For example, the lowest energy state of helium, in which two electrons are placed in the $1s$ orbital, is written as $1s^2$. (This is to be read as one-s-two and not as one-s-squared.) When one of the electrons is placed in an orbital of higher energy, an "excited" configuration is obtained. An example might be $1s^12p^1$ which one electron is in the $1s$ orbital and one electron is in the $2p$ orbital. It should be emphasized that the concept of assigning an electronto an atomic orbital is a rigorous and exact concept *only* for the hydrogen atom; for the many-electron case it is an approximation.

The atomic orbital approximation may be tested by applying it to the helium atom. We have seen that the energy of a single electron moving in the attractive field of a nucleus of charge $+Ze$ is

$$E_n = \frac{-2\pi^2 me^4 Z^2}{n^2 h^2} = \frac{Z^2}{h^2} K \quad n = 1, 2, 3, \ldots\ldots$$

The energy of the two electrons in the helium atom, each considered to be independent of the other, is simply

$$E_1 = -\left(\frac{2^2}{1}\right)K - \left(\frac{2^2}{1}\right)K = -8k$$

for the $1s^2$ electron configuration. To this energy value must be added the energy of replusion between the two electrons. Since both electrons have been placed in a $1s$ atomic orbital, we know that the charge distribution for each electron must be spherical and centred on the helium nucleus. The two charge distributions will be completely intermingled and we must calculate the energy of replusion between every small element of charge density of the one distribution with every small charge element of the second distribution. This calculation can be readily done by the methods of integral calculus and the value of the average energy replusion is found to be

$$E_c = \frac{5}{4} ZK = \frac{5}{2} K \qquad (\angle\text{He} = 2)$$

We label this energy E_c as it is a correction to our first approximation to the energy. Notice that in general E_c depends directly on the value of Z. This makes physical sense, for the greater the value of Z, the more contracted and superimposed are the two charge distributions, and the greater is the energy of replusion between them. Note as well that the correction E_c is indeed smaller than E_1, an assumption we made in developing this method of approximating the electronic energy.

The estimate of the total electronic energy of helium atom is

$$E_{He} = E_1 + E_c$$

$$= -8K + \frac{5}{2}K$$

$$= -\frac{11}{2}K$$

This total energy is called the electronic binding energy as it is the energy released when two initially free electrons are bound to the helium nucleus Recall that $-K$ represents the binding energy of the most stable state of the hydrogen atom; thus the helium atom is five and one half times more stable than the hydrogen atom. This is not really a fair comparison because the value $(11/2)K$ is the energy required to remove both electrons from the helium atom (a double ionization)

$$He \rightarrow He^{+2} + 2e^- \qquad \Delta E = \frac{11}{2}K$$

It is more interesting to compare the energy required to remove a single electron from helium with the energy required to remove the single electron in hydrogen. The energy of the reaction (The energy required to ionize an atom will be denoted by the letter I.)

$$He \rightarrow He^+ + e^- \qquad \Delta E = I$$

is easily caluclated

$$I = E_{He} + E_{e^-} - E_{He}$$

The energy of the He^+ ion using equation with $n = 1$ and $Z = 2$, is

$$E_{He^+} = -4K$$

as there is but a single electron left. The energy of the ionized electron E_e- is set equal to zero as it is assumed to be at rest infinitely far away from the ion and $E_{He} = -(11/2)K$. Thus I, the first ionization potential of helium, is equal to 1.5K, or one and one half times larger than the energy required to ionize a hydrogen atom.

How well do our calculated values for the ionization potential and total energy agree with the experimental results? The energy required for the removal of both electrons is 78.98 ev. Since K = 13.61 ev, the calculated value is (11/2)13.61 = 74.86 ev. This is an encouraging result as the error is only about 5 per cent. The experimental value for I is 24.58 ev and our calculated value is (1.5)13.61 = 20.42 ev. The percentage error is larger for the latter case because the actual error is the same in both calculations but I is smaller than ΔE. However, the method seems promising. We have indeed predicted that it requires almost twice as much energy to remove an electron from helium as it does to remove one from hydrogen.

The calculations outlined above may be improved by introducing the concept of an *effective nuclear charge*. Since there are two electrons present in the helium atom, neither electron experiences the full attractive force of the two positive charges on the helium nucleus. Each electron partially screens the nuclear charge from the other. We saw previously that the averagevalue of the distance between an electron and the nucleus for a strictly hydrogen-like orbital varied as $(1/Z)$. Thus by assuming that each electron moves in the field of the full nuclear charge of helium, we consider it to be in a 1s orbital with exactly one half the value of $\bar{r}$ as that found for a hydrogen 1s orbital. Since the electron on the average experiences a rereduced nuclear charge (i.e., the effective nuclear charge) because of the screening effect of the second electron, we should place it in a 1s orbital which possesses an $\bar{r}$ value somewhere between that found for an orbital for the cases Z = 1 and Z = 2. In other words, the size of the orbital should be determined by an effective nuclear charge, rather than by the actual nuclear charge. This lowered value for Z will obviously decrease the value of the average repulsion energy between the electrons as the two charge clouds will be more expanded and the average distance between the charge points in each distribution will increase. An increased value of $\bar{r}$ will also decrease the average kinetic energy of the electrons and thus again lead to an increase in the stability of the atom. On the other hand, an increase in $\bar{r}$ will lead to a less negative potential energy as the electrons will on the average be further away from the nucleus. Thus there is some best value for the effective nuclear charge and for $\bar{r}$, the value which gives the most stable description of the atom. For helium this "best" value for the effective nuclear charge is found to be 1.687 and the total energy of He is now calculated to be 77.48 ev. The error has been reduced to approximately 2 per cent.

The effective nuclear charge value cannot be inserted into equation to determine the energy of the electron. The Z in equation refers to the actual nuclear charge, while the effective nuclear charge is a number, always less than the actual Z. which determines the optimum size of the orbital when other electrons are present. The value of Z appearing in the equate for E_c will be the effective nuclear charge value. The value of E_c is indeed determined solely by the degree to which the two electron distributions are contracted and this is governed by the effective nuclear charge. It should be pointed out that the concept of an effective nuclear charge will be paramount in our future discussions concerning the electronic structures and properties of many-electron atoms.

Excited states of the helium atom. Just as the single electron in the hydrogen atom can be excited to higher quantum levels, so it should be possible to excite one of the electrons in the He atom to energy levels with quantum numbers greater than one. This will change the electron configuration from $1s^2$ to say, $ls^1 2s^1$ or $1s^1 2p^1$ etc. The excited electron may again lose the excitation energy in the form of light and fall back to the $1s$ level, giving the ground electronic configuration $1s^2$

$$Is^1 2p^1 \rightarrow 1s^2 + hv$$

Thus the helium atom should emit a line spectrum when it is excited in an electrical discharge tube. Since only a single electron is excited at a given time although if is possible with the use of a laser to excite two electrons simultanously-the spectrum for helium should be formally the same as that observed for hydrogen. However, since the nuclear charge experienced by the electron will always be greater than one, the lines in the helium spectrum should be observed at higher frequencies (shorter wavelengths) than those for hydrogen.

Table. The Wavelengths for the Balmer Series in H and the Wavelengths for the Corresponding One-Electron Transitions in He

H		He	
n	λ (Å)	n	λ (Å)
3	6563	6	5016
4	4861	4	3965
5	4340	5	3614
6	4101	6	3449

In Table 4-1. we compare two corresponding line spectra, one for hydrogen and one from helium. In both cases the excited electron fall from an upper *p* energy level to the 2*s* energy level. In hydrogen the frequencies of the lines in the spectrum are determined by the energy differences between the configuration.

$$np^1 \rightarrow 2s^1 \qquad n = 3, 4, 5, 6, \ldots\ldots$$

This series of jumps from E_n (n = 3, 4, 5, 6, . . .) to E_2 level generates the Balmer series of lines which we discussed earlier. We are now being more specific in stating that in this particular examples the excited electrons in in an *n*p orbital. The helium spectrum we wish to compare with this one arises from the transition between configurations

$$1s^1np^1 \rightarrow 1s^1 2s^1 \qquad n = 3, 4, 5, 6, \ldots\ldots$$

Qualitatively the two spectra are the same, as our model predicted. In addition, the helium lines occur at shorter wavelengths (higher energies) than for hydrogen. In fact, for every series of lines (Lyman, Balmer, etc.) found for hydrogen, there is a corresponding series found at shorter wavelengths for helium. Our model, which uses hydrogen-like atomic orbitals to describe many-electron atoms, looks promising indeed. However, it is in the study of the spectrum of helium that we encounter the first shortcoming of this simple approach; there are two series of lines observed for helium for every single series of lines observed for hydrogen. Not only does helium possess the "Balmer" series quoted in Table 4-1, it has a second "Balmer"series starting at $\lambda = 3889$A. That is, the whole series is repeated at shorter wavelengths. Rather than abandon the atomic orbital approach for the many-electron atom, let us keep the above failure of the method in mind and proceed with an application to the lithium atom.

The lithium atom. There are three electrons in the lithium atom ($Z = 3$) but the total repulsion energy between the electrons is still determined by considering the repulsions between a pair of electrons at a time. For this reason, three electrons are fundamentally no more difficult to treat than two electrons. There are simply more possible pairs and hence more repulsive interactions to consider than in the two-electron case. The dependence of the potential energy on the distances between the electrons and between the electrons and the nucleus is

$$PE = -\frac{3e^2}{r_1} - \frac{3e^2}{r_2} - \frac{3e^2}{r_3} + \frac{e^2}{r_{12}} + \frac{e^2}{r_{12}} + \frac{e^2}{r_{23}}$$

where r_{12}, is the distance between electrons 1 and 2, r_{13} the distance between electrons 1 and 3, and r_{23} the distance between electrons 2 and 3.

It is natural to assume that, as in the case of hydrogen or helium, the most stable energy of the lithium atom will be obtained when all three electrons are placed in the 1*s* atomic orbital giving the electronic configuration $1s^3$. Proceeding as in the case of helium we calculate the first approximation to the energy to be

$$E_1 = -3\left(3^2 K\right) = -27K$$

This represents the sum of the energies obtained when each electron is considered to move independently in the field of the nucleus of charge +3 in an orbital with $n = 1$. To this must be added the energy of repulsion between the electrons. The average repulsion energy between a pair of electrons is again given by $\frac{5}{4}ZK$. In lithium we must consider the repulsion between electrons 1 and 2, between electrons 1 and 3, and between electrons 2 and 3. Therefore the total repulsion energy which represents the correction to E is

$$E_c = 3\left(\frac{5}{4}ZK\right)$$
$$= \frac{45}{4}K$$

and the total electronic energy of the lithium atom is predicted to be

$$E = E_1 + E_c$$
$$= -27K + \frac{45}{4}K$$
$$= -15.75K$$
$$= -214.4ev$$

Thus it should require 214.4 ev to remove all three electrons from the lithium atom.

We can also calculate the energy required to remove a single electron from lithium (the first ionization potential)

$$Li \rightarrow Li^+ + 1e^- \qquad \Delta E = I$$

$$I = E\left(Li^+\right) - E(Li)$$

$$I = \underbrace{-2\left(3^2 K\right) + \left(\frac{15}{4}K\right)}_{E\left(Li^+\right)} - \underbrace{(-15.75K)}_{E(Li)}$$

$$= +1.50K \quad = +20.4ev$$

When the predicted values for lithium are compared with the corresponding experimental values, they are found to be in serious error. The lithium atom is not as stable as the calculations would suggest. Experimentally it requires 202.5 ev to remove all three electrons from lithium, and only 5.4 ev to remove one electron. Experimentally it requires less energy to ionize lithium than it does to ionize hydrogen, yet our calculation predicts an ionization energy one and one half times larger. The error in I is 300 per cent! We should expect a realistic model to do better than this. More serious than this, however, is that the kind of calculation we are doing should never predict the system to be more stable than it actually is. The method should always predict an energy less negative than is actually observed. If this is not found to be the case, then it means that an incorrect assumption has been made or that some physical principle has been ignored. It is also clear that if we were to continue this scheme of placing each succeeding electron in the 1*s* orbital as we increased the nuclear charge by unity we would never predict the most striking property of the elements: the property of periodicity.

We might recall at this point that there is a periodicity in the types of atomic orbitals found for the hydrogen atom. With every increase in n, all the preceding values of l are repeated, and a new l value is introduced. If we could discover a physical reason for not placing all of the electrons in the1*s* orbital, but instead place succeeding electrons in the orbitals with higher n values, we could expect to obtain a periodicity in our predicted electronic structures for the atoms. This periodicity in electronic structure would then explain the observed periodicity in their properties. There must be another factor governing the behaviour of electrons and this factor must be one which determines the number of electrons that may be accommodated in a given orbital. To discover what this missing factor is and to find the physical basis for it, we must investigate further the magnetic properties of electrons.

THE MAGNETIC PROPERTIES OF THE ELECTRON

So far, the only motion we have considered for the electron is a motion in three-dimensional space. Since this motion is ultimately described in terms of an orbital wave function, we term this the orbital motion of the electron. However, the electron may possess an internal motion of some kind, one which is independent of its motion through space. Since the electron bears a charge, such an internal motion, if it does exist, might be expected to generate a magnetic moment. We have previously pointed out that when an electron is in an atomic orbital for which l is not equal to zero, the resultant angular motion of the electron gives rise to a magnetic moment. We would anticipate then that an electron in an s orbital ($l = 0$) should not exhibit any magnetic effects as its angular momentum is zero. If an electron in these circumstances did exhibit a magnetic effect, it would indicate that another type of motion was possible, presumably an internal one. Whether or not an electron in an s orbital does possess a magnetic moment may be determined by means of an atomic beam experiment similar to the one previously described.

In the present experiment a beam of hydrogen atoms is passed through the apparatus. All of the hydrogen atoms in the beam will be in their ground state with $l = 0$ and hence they will not possess an orbital magnetic moment. However, when the magnetic field is applied, something does happen to the beam of atoms. It is split into two distinct beams, one of which is deflected to the N pole of the magnet and the other to the S pole. Thus even when atoms possess no magnetic moment because of the orbital motion of the electrons, they may still exhibit magnetic effects! As striking as the behaviour of the atoms as small magnets is the splitting of the beam into two distinct components. Let us consider first the origin of the magnetic effect, and second, the splitting of the beam into two distinct beams.

The observed magnetism of the hydrogen atoms must be due to some motion of the electrons. The nucleus of a hydrogen atom does possess a magnetic moment but its magnitude is too small, by a factor of roughly a thousand, to account for the deflections observed in this experiment. A magnetic moment will be observed only when the charged particle possesses angular momentum. Since the orbital angular momentum for an electron in the ground state of hydrogen is zero, we are forced to assume that the electron possesses some internal motion which has associated with it an angular momentum. A classical analogue of the internal angular momentum would be a spinning motion of the electron about its own axis. For this reason it is referred to as a *spin angular momentum* and the associated magnetic effect as a *spin magnetic moment.* These effects are separate from, and in addition to, the orbital angular momentum of the electron (classically, the rotation of the electron around the nucleus) and its associated magnetic effects.

We are familiar enough with the predictions of quantum mechanics to anticipate that the spin angular momentum and its component along some axis will be quantized. As in the case of orbital angular momentum, the effect of the quantization will be to limit the number of values which the component of the spin magnetic moment may have along any given axis. The magnitude of the spin angular momentum will determine the number of possible values its component may have along a given axis. Each of the possible values will in turn cause some fraction of the total spin magnetic moment to be aligned along the same axis. In the case of the electron's orbital motion, we found that as l and hence the orbital angular momentum was increased, the number of possible values for the component of the orbital magnetic moment along a given axis was increased, the number being equal to $(2l + 1)$.

We can use a magnetic field to inquire into the nature of the spin angular momentum as well. In fact, we have already discussed the pertinent experiment. The beam of hydrogen atoms was split into just two components in the atomic beam experiment. This means that the component of the electron's spin magnetic moment (and spin angular momentum) along a given axis may have only one of two possible values; the component may be aligned with the field and hence be attracted, or it may be opposed to the field and be repelled. The electron's spin magnetic moment has been detected in many different kinds of experiments and the results are remarkable in that only two components of constant magnitude are ever observed. The electron is always either repelled by the field or attracted to it. This implies that the magnitude of the spin angular momentum for a single electron may have *only one possible value.* Since the number of possible values for the component of a given amount of angular momentum of any type in quantum mechanics is $(2l + 1)$, l must equal ½ and only ½ for the spin angular momentum, and the values of m for the electron spin, which assume values from a maximum of l to a minimum of $-l$ in steps of unity, must equal +½ and –½. In this respect the spin angular momentum of the electron is quite different from its orbital angular momentum, which may have many possible values, as the value of l for the orbital motion is restricted only in that it must equal zero or an integer.

It should be stressed that the splitting of the beam of hydrogen atoms into only two components is again evidence of quantization. If the atomic magnets (the hydrogen atoms) behaved according to classical mechanics, then the effect of the magnetic field would be simply to broaden the beam. The orientations of the atomic magnets would be random when they first entered the field of the magnet and classically the individual atomic magnets could be aligned at any and all angles with respect to the field, giving all possible components of the spin magnetic moment along the direction of the field. The inhomogeneous field would then exert a force proportional to the magnitude of the component, and the beam would broaden but not split.

Since the spin magnetic moment is an intrinsic property of the electron, even a beam of free electrons should be split into two components in a magnetic field. However, the charge possessed by the free electron also interacts with the magnetic field and the much smaller magnetic-magnetic interaction is masked by the usual deflection of a charge species in a magnetic field. By employing a neutral atom, the complications of the electronic charge may be avoided.

Let us summarize what we have learned about this new property of the electron. Since an electron may exhibit a magnetic moment even when it does not possess orbital angular momentum, it must possess some internal motion. We call this motion the electron spin and treat it quantum mechanically as another kind of angular momentum. Experimentally, however, all we know is that the electron possesses an intrinsic magnetic moment. The remarkable feature of this intrinsic magnetic moment is that its magnitude and the number of components along a given axis are fixed. A given electron may exhibit only one of two possible components; it may be aligned with the field or against it. Experimentally, or theoretically, this is all we can know about the spin magnetic moment and the spin angular momentum. Hence only one quantum number is required to describe completely the spin properties of a single electron. We shall denote the value of this quantum number by ↑ or ↓, the upwards-pointing arrow signifying that the component of the magnetic moment is aligned with the field and the downwards-pointing arrow that this component is opposed to the field.

A total of four quantum numbers is required to specify completely the state of an electron when it is bound to an atom. The quantum numbers *n*, *l* and *m* determine its energy, orbital angular momentum and its component of orbital angular momentum. The fourth quantum number, the spin quantum number, summarizes all that can be known about the spin angular momentum of the electron. This final quantum number may have only one of two possible values corresponding to the magnetic moment component being (a) aligned with the field or (b) opposed to it.

THE PAULI EXCLUSION PRINCIPLE

The consequences of the spin quantum number, when applied to the problem of the electronic structure of atoms, are not immediately obvious. The small magnitude of the electron's magnetic moment does not directly affect the energy of the electron to any significant degree. To see just how the spin of the electron does influence the problem, let us reconsider our atomic orbital model in the light of this new degree of freedom for the electron. In particular let us reconsider those instances in which our model failed to account for the observations.

If a beam of helium atoms is passed through a magnetic field, no splitting and no deflection is observed. The helium atom, unlike the hydrogen atom is not magnetic. We could account for the absence of a magnetic moment for helium if we assumed that of the two electrons in the helium 1*s* orbital, one had its magnetic moment component up (↑) and the other down (↓). The two components would then cancel and there would be no resultant magnetic effect. Our complete description of the electronic configuration of the helium atom would be $1s^2(\uparrow\downarrow)$, i.e., both electrons have $n = 1$, $l = 0$, $m = 0$ and one has a spin ($\uparrow$) and the other a spin ($\downarrow$).

You may wonder why the states of helium corresponding to the configurations $1s^2(\uparrow\uparrow)$ or $1s^2(\downarrow\downarrow)$ are not observed. These states should exhibit twice the magnetism possessed by a hydrogen atom. They are, however *not found to occur*. What about the excited states of the helium atom? An excited state results when one electron is raised in energy to an orbital with a higher n value. The electrons are thus in different orbitals. The spin assignments for an excited configuration can be made in more than one way and are such as to predict the occurrence of both magnetic and non-magnetic helium. For example, the configuration $1s^12s^1$ could be $1s^1(\uparrow)2s^1(\downarrow)$ and be nonmagnetic or it could equally well be $1s^1(\uparrow)2s^1(\uparrow)$ and be magnetic.

Both the magnetic and non-magnetic forms are indeed found to occur for helium in an excited state. There are in fact two kinds of excited helium atoms, those which are non-magnetic and those which are magnetic. If the two forms of helium possess different energies even though they have the same orbital configuration (we shall see why this should be so later) then we have an explanation for the previously noted discrepancy that helium exhibits twice the number of line spectra as does hydrogen. For every set of lines in the spectrum which arises from the transition of the electron from the configurations $1s^1(\uparrow)np^1(\downarrow)$ to the configuration $1s^1(\uparrow)2s^1(\downarrow)$ for example, there will be another set of lines due to transitions from $1s^1(\uparrow)np^1(\uparrow)$ to $1s^1(\uparrow)2s^1(\uparrow)$.

The study of the magnetic properties of the ground and excited states of helium is sufficient to point out a general principle. For the ground state of helium, in which both electrons are in the same atomic orbital, *only the non-magnetic form exists*. This would imply that when two electrons are in the same atomic orbital their spins must be paired, that is, one up ($\uparrow$) and one down ($\downarrow$). This is *an experimental fact* because helium is never found to be magnetic when it is in its electronic ground state. When the electrons are in different orbitals, then it is again an experimental fact that their spins may now be either paired (↓↑) or unpaired, e.g., (↑↑). Thus when two electrons are in the same orbital (i.e., they possess the same *n*, *l* and *m* values) their spins must be paired. When they are in different orbitals (one or more of their *n*, *l* and *m* values are different) then their spins may be paired or unpaired. We could generalize these observations by stating that "*no two electrons in the same atom may have all four quantum numbers the same.*" Stated in this way we see immediately that any given orbital may hold no more than two electrons. Since two electrons in the same orbital have the same values of *n*, *l* and *m*, they can differ only through their spin quantum number. However, the spin quantum number may have only one of two possible values, and these possibilities are given by (*n*, *l*, *m*, ↑) or (*n*, *l*, *m*, ↓).

We have indeed found the principle we were seeking, one which limits the occupation of an atomic orbital. This principle is known as the *Pauli exclusion principle*. One form of it, suitable for use within the framework of the orbital approximation, is the statement given in quotation marks above. The Pauli principle cannot be derived from, nor is it predicted by, quantum mechanics. It is a law of nature which must be taken into account along with quantum mechanics if the properties of matter are to be correctly described. The concept of atomic orbitals, as derived from quantum mechanics, together with the Pauli exclusion principle which limits the occupation of a given orbital, provides an understanding of the electronic structure of many-electron atoms. We shall demonstrate this by "predicting" the existence of the periodic table.

THE ELECTRONIC BASIS OF THE PERIODIC TABLE

The hydrogen-like orbitals for a many-electron atom are listed in order of increasing energy in Fig.. This energy level diagram differs from the corresponding diagram for the hydrogen atom, a one-electron system. In the many-electron atom all orbitals with the same value of the principal quantum number *n* do not have the same energy as they do in the case of hydrogen. For the many-electron atoms, the energy of an orbital depends on both *n* and *l*, the energy increasing as *l* increases even when *n* is constant. For example, from Fig it is evident that the 3*d* orbital possesses a higher energy than does the 3*p* orbital which in turn has a higher energy than does the 3*p* orbital. The reason for this difference between the one- and the many-electron case will be discussed below. The energy of the orbital is still independent of the magnetic quantum number *m*. Thus when $l = 1$, there are three *p* orbitals which are still degenerate (all possess the same energy) and this is indicated by the three open circles which are superimposed on each of the *p* levels. The open circles thus represent the number of available orbitals or the degeneracy of each orbital energy level.

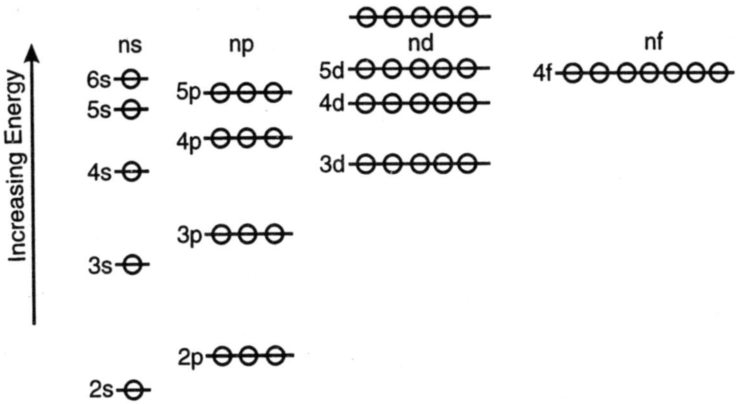

Fig. An Orbital Energy Level diagram for a Many-Electron Atom.

With the aid of this energy level scheme and the Pauli principle we may proceed to build up the electronic structures of all the atoms. We do this by asssigning electrons one at a time to the vacant orbital which possesses the lowest energy. An orbital is "filled" when it contains two electrons with their spins paired.

Hydrogen. The nuclear charge is 1 and the single electron is placed in the 1*s* orbital. The electronic configuration is $1s^1$.

Helium. The nuclear charge is increased by one unit to 2 and the extra electron is again placed in the 1*s* orbital, with its spin opposed to that of the electron already present. The electronic configuration is $1s^2$.

Lithium. The nuclear charge is 3 and the third electron, because of the Pauli principle, must be placed in the 2s orbital as the 1s orbital is doubly occupied. The electronic configuration of lithium is therefore $1s^2 2s^1$.

We can now answer the question as to why the 2*s* orbital is more stable than the 2*p* orbital, i.e., why Li is described as $1s^2 2s^1$ and not as $1s^2 2p^1$. The two inner electrons of lithium (those in the 1*s* orbital) partially shield the nuclear charge from the outer elctron. Recall that as *n* increases, the average distance between the electron and the nucleus increases. Thus most of tthe electron density of the electron with $n = 2$ will lie outside of the charge density of the two inner electrons which have $n = 1$. When the outer electron is at large distances from the nucleus and thus essentially outside of the inner shell of electron density it will experience a force from only one of the three positive charges on the lithium nucleus. However, as the outer electron does have a small but finite probability of being close to the nucleus, it will penetrate to some extent the tightly bound electron density of the two 1*s* electrons. In doing so it will "see" much more of the total nuclear charge and be more tightly bound. The closer the outer electron can get to the nucleus, that is, the more it can penetrate the density distribution of the inner shell electrons, the more tightly bound it will be.

An electron in an *s* orbital has a finite probability of being found right at the nucleus. An electron in a *p* orbital on the other hand has a node in its density distribution at the nucleus. Thus an *s* electron penetrates the inner shell density more effectively than does a *p* electron and is consequently more tightly bound to the atom. In a hydrogen atom, there are no inner electrons and both a 2*s* and 2*p* electron always experience the full nuclear charge and have the same energy. The crux of this penetration effect on the energy is that the inner shell electron density does possess a finite extension in space. Thus an outer electron can penetrate inner shell density and the screening effect is reduced. If the inner shell density was contracted right onto the nucleus, then no matter how close the outer electron came to the lithium nucleus, it would always experience only a charge of +1. This dependence of the orbital energies on their *l* value is aptly called the penetration effect.

The electron density of a *d* electron is concentrated even further away from the nucleus than is that of a *p* electron. Consequently, the orbital energy of a *d* electron is even less stable than that of a *p* electron. In some atoms the penetration of the inner shell density by a *d* electron is so slight that its energy is raised even over that of the *s* electron with the next highest *n* value. For example, a 3*d* electron possesses a higher energy than does a 4*s* electron in the atoms *Sc* to *Zn* with the exceptions of *Cr* and *Cu*. The penetration effect in these elements overrides the principal quantum number for *d* electrons in determining their relative energies.

Notice that the configuration $1s^2 2s^1$ for lithium overcomes the difficulties of our earlier attempts to describe the electronic structure of this atom. The Pauli principle, of which we were ignorant in our previous attempt, forces the third electron to occupy the 2*s* orbital and forces in turn the beginning of a new quantum shell, that is, a new value of *n*. Thus lithium, like hydrogen, possesses one outer electron in an *s* orbital. Since it is only the outer electron density which in general is involved in a chemical change, lithium and hydrogen should have some chemical properties in common, as indeed they do. Hydrogen is the beginning of the first period ($n = 1$) and lithium marks the beginning of the second period ($n = 2$).

Beryllium. The nuclear charge is 4 and the electronic configuration is $1s^2 2s^2$.

Boron. $Z = 5$ and the electron configuration is $1s^2 2s^2 2p^1$.

Carbon. $Z = 6$. The placing of the sixth electron of carbon requires some comment. It will obviously go into a 2*p* orbital. But in which of the three should it be placed? Should it be placed in the 2*p* orbital which already possesses one electron, or should it be placed in one of the vacant 2*p* orbitals? If it is placed in the occupied 2*p* orbital its spin must be paired with that of the electron already present and the result would be a nonmagnetic carbon atom. If, however, it is placed in one of the vacant 2*p* orbitals it may be assigned a spin parallel to the first electron. The question is decided on the grounds of which situation gives the lowest energy. As a result of the Pauli principle, two electrons with parallel spins (both up or both down) have only a very small probability of being close to one another. In fact the wave function which describes the two-electron case for parallel spins vanishes when both electrons approach one another. When the wave function vanishes, the corresponding probability distribution goes to zero. On the average, then, electrons with parallel spins tend to keep out of each other's way. Two electrons with paired spins, whether in the same or different orbitals are not prevented by the Pauli principle from being close to one another. The wave function for this situation is finite even when they are on top of one another! Obviously, two electrons with parallel spins will have a smaller value for the electrostatic energy of repulsion between them than will two electrons with paired spins. This is a general result which holds almost without exception in the orbital approximation. It is known as one of Hund's rules as he was the first to state it. Thus the most stable electronic configuration of the carbon atom is $1s^22s^22p^2(\uparrow\uparrow)$ where we have emphasized the fact that the two $2p$ electrons have parallel spins and hence must be in different $2p$ orbitals.

Nitrogen. $Z = 7$. Because of Hund's rule the electronic configuration is

$$1s^22s^22p^3(\uparrow\uparrow\uparrow)$$

i.e., one electron in each of the $2p$ orbitals. The configuration with the largest possible component of the spin magnetic moment will be the most stable.

Oxygen. $Z = 8$. One of the $2p$ electrons must now be paired with the added electron, but the other $2p$ electrons will be left unpaired.

$$1s^22s^22p^4(\uparrow\uparrow)$$

(Only the number of unpaired electrons is indicated by the arrows.)

Fluorine. $Z = 9$. The configuration will be

$$1s^22s^22p^5(\uparrow)$$

Neon. $Z = 10$. The tenth electron will occupy the last remaining vacancy in the second quantum shell (the set of orbitals with $n = 2$).

$$1s^22s^22p^6$$

Thus neon represents the end of the second period and all the electrons have paired spins.

When all the orbitals in a given shell are doubly occupied, the resulting configuration is called a "closed shell." Helium and neon are similar in that they both possess closed shell configurations. Because neither of these elements possesses a vacancy in its outer shell of orbitals both are endowed with similar chemical properties. When the orbitals belonging to a given l value contain either one electron each (are half-filled) or are completely filled, the resulting density distribution is spherical in shape. Thus the electron density distributions of nitrogen and neon, for example, will be spherical.

Reference to Fig. indicates that the next element, sodium, will have its outer electron placed in the 3*s* orbital and it will be the first element in the third period. Since its outer electronic structure is similar to that of the preceding elements, lithium and hydrogen, it is placed beneath them in the periodic table. It is obvious that in passing from sodium to argon, all of the preceding outer electronic configurations found in the second period ($n = 2$) will be duplicated by the elements of the third period by filling the 3*s* and 3*p* orbitals. For example, the electronic structure of phosphorus ($Z = 15$) will be

$$1s^2 2s^2 2p^6 3s^2 3p^3 (\uparrow\uparrow\uparrow)$$

and thus resemble nitrogen.

Argon. $Z = 18$. Argon will have filled 3*s* and 3*p* orbitals and will represent the end of a period. Argon, like helium and neon, possesses a closed shell structure and is placed beneath these two elements in the periodic table.

The Transition Elements. The beginning of the fourth period will be marked by the single and double occupation of the 4*s* orbital to give potassium and calcium respectively. However, reference to the orbital energy level diagram indicates that the 3*d* orbital is more stable than the 4*p* orbital. Since there are five *d* orbitals they may hold a total of ten electrons. Thus the ten elements beginning with scandium ($Z = 21$) will possess electronic structures which differ from any preceding them as they are the first elements to fill the *d* orbitals. A typical electronic configuration of one of these elements is that of manganese: $[Ar]4s^2 3d^5$. The symbol [Ar] is an abbreviated way of indicating that the inner shells of electrons on manganese possess the same orbital configuration as those of argon. In addition, the symbol $3d^5$ indicates that there are five electrons in the $3d$ orbitals, no distinction being made between the five different d orbitals. This series of elements in which the $3d$ orbitals are filled is called the *first transition series*. The element zinc with a configuration $[Ar]4s^23d^{10}$ marks the end of this series. The six elements from gallium to krypton mark the filling of the $4p$ orbitals and the closing, with krypton, of the fourth quantum shell and the fourth period of the table.

While the $3d$ orbitals are less stable than the $4s$ orbitals in the neutral atom (with the exceptions of *Cr* and *Cu*) and are filled only after the $4s$ orbitals are filled, the relative stability of the $4s$ and $3d$ orbitals is reversed in the ionic forms of the transition metals. For example, the configuration of the ion which results when the manganese atom loses two electrons is Mn^{+2} $[Ar]3d^5$ and not $[Ar]4s^23d^3$. This is a general result. The d orbitals of quantum number n are filled only after the s orbital of quantum number $(n + 1)$ is filled in the neutral atom, but the nd orbital is more stable than the $(n + 1)s$ orbital in the corresponding ion.

The fifth period begins with the filling of the $5s$ orbital, followed by the filling of the $4d$ orbitals, which generates the second transition series of elements. The period closes with the filling of the $5p$ orbitals and ends with xenon.

The lanthanide and actinide elements. The sixth period is started by the filling of the $6s$ orbital. The next element, lanthanum, has the electronic configuration $[Xe]6s^25d^1$. However, the next fourteen elements represent the beginning of another new series as the filling of the $4f$ orbitals is now energetically favoured over a continued increase in the population of the $5d$ orbitals. Note that the very small penetration effect possessed by the $4f$ orbitals (n = 4) delays their appearance until the sixth quantum shell has been partially filled. There are fourteen elements in this series, called the *lanthanide series*, since there are seven $4f$ orbitals ($l = 3$ and $2 \times 3+1 = 7$).

The third transition series follows the lanthanide elements as the occupation of the $5d$ orbitals is completed. This in turn is followed by the filling of the $6p$ orbitals. The final period begins with the filling of the 7*s* orbital and continues with the filling of the 5*f* orbitals. This second series of elements with electrons in *f* orbitals is called the *actinide series.*

The concept of atomic orbitals in conjunction with the Pauli principle has indeed predicted a *periodicity in the electronic structures of the elements*. The form of this periodicity duplicates exactly that found in the periodic table of the elements in which the periodicity is founded on the observed chemical and physical properties of the elements. Our next task will be to determine whether or not our proposed electronic structures will properly predict and explain the observed variations in the chemical and physical properties of the elements.

Electronic Basis for the Properties of the Elements

We shall now present an interpretation of the physical and chemical properties of the elements based on the atomic orbital description of their electronic structures. Our discussion of the properties of the atoms will be a qualitative one, but it should be pointed out that many of the properties of atoms can now be accurately predicted by quantum mechanical calculations employing a very extended version of the atomic orbital concept.

HORIZONTAL VARIATIONS

The experimental values of the atomic radii and the first and second ionization potentials of the elements (labelled as I_1 and I_2 respectively) in the third row of the periodic table are listed in Table. A study of these values will indicate the basic trends observed as the number of electrons is increased one at a time until all the orbitals with a given value of *n* are fully occupied.

Table. The Atomic Radii and Ionization Potentials* of Third Row Elements

Element	Na	Mg	Al	Si	P	S	Cl	Ar
Radius (Å)	1.86	1.60	1.48	1.17	1.0	1.06	0.97	
I_1 (ev)	5.14	7.64	5.98	8.15	11.0	10.4	13.0	15.8
I_2 (ev)	47.3	15.0	18.8	16.3	19.7	23.4	23.8	27.6

Atomic Radii

The diameter of an atom is difficult to define precisely as the density distribution tails off at large distances. However, there is a limit as to how close two atoms can be pushed together in a solid material. We shall take one half of the distance between the nuclei of two atoms in an elemental solid as a rough measure of the atomic radius. Any consistent method of defining the radius leads to the same trend we see in Table. The size of the atom in general decreases as the number of electrons in the quantum shell is increased. This observation, which at first sight might appear surprising, finds a ready explanation through the concept of an *effective nuclear charge.*

The electric field and hence the attractive force exerted by the nucleus on an electron in the outer quantum shell is reduced because of the screening effect of the other electrons which are present in the atom. An outer electron does not penetrate to any great extent the tightly bound density distribution of the inner shell electrons. Consequently each inner electron (an electron with an n value less than the n value of the electron in question) reduces the value of the nuclear charge experienced by the outer electron by almost one unit. The remaining outer electrons on the other hand are, on the average, all at the same distance away from the nucleus as is the electron under consideration. Consequently each outer electron screens considerably less than one nuclear charge from the other outer electrons. Thus the higher the ratio of outer shell to inner shell electrons, the larger will be the "effective nuclear charge" which is experienced by an electron in the outer shell.

All of the elements in a given row of the periodic table possess the same number of inner shell electrons. For example, the elements in the third row have the inner shell configuration of $1s^2 2s^2 2p^6$. As we move across the periodic table from left to right the nuclear charge increases, and each added electron is placed in the outer shell until a total of eight is reached and the quantum shell is full. The number of outer shell electrons increases along a given period, but the number of inner shell electrons remains fixed. Thus the effective nuclear charge increases from a minimum value for sodium, where the ratio of outer shell to inner shell electrons is

1:10, to a maximum value for argon where the same ratio is 8:10. The atomic radius undergoes a gradual decrease since the outer electrons become more tightly bound as the effective nuclear charge increases.

These features of the atomic density distributions are clearly evident in a graph of the radial distribution function, $Q(r)$. This function, it will be recalled, gives the number of electronic charges within a thin shell of space lying between two concentric spheres, one of radius r and the other with a radius only slightly larger. The radial distribution functions for atoms may be determined experimentally by X-ray or electron diffraction techniques.

Plots of $Q(r)$ versus r for sodium and argon, the first and last members of the third row of the periodic table, clearly reveal the persistence of a "shell structure" in the many-electron atoms.

There are three peaks in the density distribution corresponding to the presence of three principal quantum shells in the orbital model of the electronic structure of sodium and argon. The peak closest to the nucleus may be identified with the charge density in the 1*s* orbital, the middle peak with that in the 2*s* and 2*p* orbitals and the outer peak with the charge density in the 3*s* orbital in sodium and in the 3*s* and 3*p* orbitals in argon. The maxima in $Q(r)$ occur at smaller values of r for argon than for sodium as expected on the basis of a larger effective nuclear charge for argon than for sodium. Most of the 1*s* charge density is found within a very thin shell close to the nucleus in both cases as the inner shell density experiences the field of the full nuclear charge, $Z_{Na} = 11$ and $Z_{Ar} = 18$. The charge density in the $n = 2$ orbitals is confined to a shell which is narrower and closer to the nucleus in argon than in sodium. The electrons in this second shell experience a nuclear charge of approximately sixteen in argon but of only nine in sodium.

The most dramatic effect of the difference in the effective nuclear charges of argon and sodium is evidenced by the appearance of the electron density in the valence shell. In sodium this shell is broad and diffuse as there are ten inner electrons shielding eleven nuclear charges. In argon where there are ten inner electrons to shield eighteen nuclear charges the valence shell is more contracted and it peaks at roughly one third of the corresponding distance in sodium. The valence shell density is clearly more tightly bound in argon than in sodium.

The effect of an increase in the nuclear charge on the individual atomic orbital densities for elements in the same row of the periodic table, in this case sodium and chlorine. The total density distribution for the atom is obtained by summing the individual orbital densities. The summation of just the 1*s*, 2*s* and three 2*p* densities yields the spherical inner shell densities indicated on the diagram as "core densities." It is the core density which shields the nuclear charge from the valence electrons. The outer density contour indicated for the inner shell or core densities defines a volume in space containing over 99 per cent of the electronic charge of the inner shell electrons. Thus the effective nuclear charge experienced by the valence density beyond the indicated radii of the core densities is $Z_{Na}-10 = 1$ for sodium and $Z_{Cl}-10 = 7$ for chlorine. Notice that the radius of the core density is smaller for chlorine than it is for sodium and thus the attractive force exerted on the valence electrons by each of the unscreened nuclear charges will be greater in chlorine than in sodium.

There is one exception to the trend of a decrease in diameter across a given row in that phosphorus has an atomic radius slightly smaller than that of sulphur which follows it in the table. The configuration of the outer electrons in phosphorus is $3s^2 3p^3(\uparrow\uparrow\uparrow)$. Each of the *p* orbitals contains a single electron and according to Hund's rule all will have the same spin quantum number. Electrons with identical spins have smaller electron-electron repulsion energies than do electrons with paired spins, for reasons we have previously mentioned. Therefore, the larger the number of parallel spins in an atom, the smaller will be the average energy of repulsion between the electrons. Three is the maximum number of unpaired spins possible in any of the short periods as this corresponds to a half-filled set of *p* orbitals. The stabilizing effect of the decreased energy of repulsion between the electrons is comparable to the effect obtained by increasing the effective nuclear charge by approximately one. This can be seen by comparing phosphorus with sulphur. Sulphur has an increased nuclear charge but the added electron must be paired up with one of the electrons in the *p* orbitals. The number of unpaired electrons with parallel spins is thus reduced to two, the average energy of repulsion between the electrons is increased, and the sulphur atom is slightly larger than the phosphorus atom.

The decrease in energy which is obtained by maximizing the number of parallel spins is not sufficient to change the most stable outer configuration actually found for silicon, $3s^2(\uparrow\downarrow)3p^2(\uparrow\uparrow)$, to that in which all four outer electrons have parallel spins, $3s^1(\uparrow)3p^3(\uparrow\uparrow\uparrow)$. This latter configuration could be obtained only by promoting an electron from a 3*s* orbital to a 3*p* orbital. The 3*s* orbital is more stable than a 3*p* orbital because of the penetration effect, and the energy increase caused by the promotion of an electron from the 3*s* to a 3*p* orbital would not be offset by the energy decrease obtained by maximizing the number of parallel spins. It is interesting to note, however, that the reverse of this is true for some of the elements in the transition series. In these elements the 4*s* and 3*d* (or in general the *ns* and (*n*–l)*d*) orbitals are the outer orbitals. The energy difference between an *ns* and an (*n*–1)*d* orbital is much less than that between an *ns* and an *np* orbital. Thus the effect of maximizing the number of parallel spins can be overriding in these cases. The outer electronic structure of vanadium is $4s^2 3d^3$. We would expect the outer electronic configuration of the next element, chromium, to be $4s^2 3d^4$ with four parallel spins. Instead, the configuration is actually $4s^1 3d^5$ resulting in a total of six parallel spins and a reduction in the energy of repulsion between the electrons.

THE IONIZATION POTENTIALS

Reference to Table indicates that in general the amount of energy required to remove one of the outer electrons increases as the effective nuclear charge increases. The increase in I_1 from approximately 5 ev for sodium to approximately 16 ev for argon dramatically illustrates the increase in the force which the nucleus exerts on the outer electrons as the nuclear charge and the number of *outer* electrons is increased. The effect of the half-filled set of *p* orbitals is again evident as I_1 is slightly larger for phosphorus than for sulphur. There is an apparent discrepancy in the value for I_1

observed for magnesium. The outer electronic configuration of magnesium is $3s^2$ and for aluminium is $3s^23p^1$. The value of 7.64 ev observed for magnesium is the energy required to remove a $3s$ electron, while the value quoted for aluminium is the energy required to remove a $3p$ electron. An s orbital is more stable than a p orbital because of its greater penetration of the inner core of electron density. Thus the penetration effect overrides the increase in the effective nuclear charge. We can test the validity of this explanation by comparing the energies required to remove a second electron (I_2) from the magnesium and aluminium atoms. The outer electronic configurations of the singly-charged magnesium and aluminium ions are $3s^1$ and $3s^2$. Thus a comparison of the second ionization potentials (I_2) will be free of the complication due to the penetration effect because we will be comparing the amount of energy required to remove an s electron in each case The values in Table indicate that the removal of an s electron requires more energy in aluminium than in magnesium, a result which is consistent with the greater effective nuclear charge for aluminium than for magnesium. What explanation can be given to the second ionization potential of sulphur being almost equal to that for chlorine?

It is worthwhile noting the large value—the largest in the table—of the second ionization potential observed for sodium. The sodium ion has the electron configuration $1s^22s^22p^6$, i.e., there are no remaining outer electrons. The second ionization potential for sodium is, therefore, a measure of the amount of energy required to remove one of what were initially inner shell electrons in the neutral atom. The effective nuclear charge experienced by a $2p$ electron in the sodium ion will be very large indeed, because the number of inner shell electrons for an $n = 2$ electron is only two. That is, only the two electrons in the $1s$ orbital exert a large screening effect. Therefore, coupled to the fact that the ion bears a net positive charge, is the fact that the ratio of outer to inner shell electrons is 8:2, which is even more favourable than that obtained for argon. The value of I_2, for sodium again emphasizes the electronic stability of a closed shell, a stability which is a direct reflection of the large value of the effective nuclear charge operative in such cases.

VERTICAL RELATIONSHIPS

Table lists the atomic radii and the ionization potentials of the elements found in the first column of the periodic table, the group I elements.

Table. Atomic Radii and Ionization Potentials of Group I Elements

Element	Li	Na	K	Rb	Cs
Radius (Å)	1.50	1.86	2.27	2.43	2.62
I_1 (ev)	5.4	5.1	4.3	4.2	3.9

The average value of the distance between the electron and the nucleus increases as the value of the principal quantum number is increased. The increase in the atomic diameters down a given group in the periodic table is thus understandable. Each of the group I elements represents the beginning of a new quantum shell. There will be a very sharp decrease in the effective nuclear charge on passing from the preceding closed shell element to a member of group I, as the number of the inner shell electrons is increased by eight. This large sudden reduction in the effective nuclear charge and the fact that the electron, because of the Pauli exclusion principle, must enter a new quantum shell, causes the group I elements to be larger in size and much more readily ionized than the preceding noble gas elements. The decrease in the effective nuclear charge and the increase in the principal quantum number down a given family bring about a steady decrease in the observed ionization potentials. Thus the outer 6*s* electron in cesium is on the average, further from the nucleus than is the outer 2*s* clcctron in lithium. It is also more readily removed.

So far we have considered the periodic variations in the energy required to remove an electron from an atom:

$$M \rightarrow M^{+} + e^{-} \qquad I_1 = \text{a positive value}$$

In some favourable cases it is possible to determine the energy released when an electron is added to an atom:

The magnitude of the energy released when an atom captures an extra electron is a measure of the atom's *electron affinity*.

It might at first seem surprising that a neutral atom may attract an extra electron. Indeed many elements do not have a detectable electron affinity. However, consider the outer electronic configuration of the group VII elements, the halogens:

$$ns^2np^5$$

There is a single vacancy in the outer set of orbitals and the effective nuclear charge experienced by the valence electrons in a halogen atom is almost the maximum value possible for any given row. Because of the incomplete screening of the nuclear charge by the outer electrons, the remaining vacancy in the outer shell will, in effect, exert an attractive force on a free electron large enough to bind it to the atom.

The electron affinities for the rare gas atoms will be effectively zero even though the effective nuclear charge is a maximum for this group of elements there are no vacancies in the outer set of orbitals in a rare gas atom and as a result of the Pauli principle, an extra electron would have to enter an orbital in the next quantum shell. The electron in this orbital will experience only a very small effective nuclear charge as all of the electrons originally present in the atom will be in inner shells with respect to it. Elements to the left of the periodic table, the alkali metals for example, do have vacancies in their outer quantum shell but their effective nuclear charges are very small in magnitude. Thus these elements do not exhibit a measurable electron affinity. The electron affinity increases across a given row of the periodic table and reaches a maximum value with the group VII elements. This is a direct reflection of the variation in the effective nuclear charge.

The orbital vacancy in which the extra electron is placed is found at larger distances from the nucleus when the principal quantum number is increased. Thus the electron affinity should decrease down any given family of elements in the periodic table. For example, the electron affinities for the halogens should decrease in the order F > Cl > Br > I.

The variation in the ionization potentials across a given row is reflected in the values shown in the atomic orbital energy level diagram for the elements from hydrogen through to neon.

The orbital energies show a uniform decrease when the nuclear charge is increased, reflecting an increase in the binding of the electrons. The total energy of a many-electron atom is *not* simply the sum of the orbital energies. Summing the orbital energies does not take proper account of the repulsions between the electrons. The orbital energies do, however, provide *approximate* estimates of the ionization potentials. The ionization potential is the energy required to remove one electron from an atom, and an orbital energy is a measure of the binding of a *single electron* in a given orbital. Thus the ionization potential should be approximately equal to *minus* the orbital energy. For example, the ionization potential of lithium is 5.39 ev and the 2*s* orbital energy is –5.34 ev. Similarly I_1, for neon is 21.56 ev and the 2*p* orbital energy is –23.14 ev.

Shell structure is also evident in the ionization potentials and orbital energies of atoms. By exposing the atom to light of very short wavelength (in the X-ray region of the spectrum), it is possible to ionize one of the inner shell electrons, rather than a valence electron. That is, the energy of an X-ray photon is comparable to the binding energy of an inner shell electron. The resulting ion is in a very unstable configuration and after a very brief period of time an electron from the outer shell "falls" into the vacancy in the inner shell. In falling from an outer to an inner shell the binding of the electron is greatly increased and a photon is emitted. The energy of this photon should be *approximately* equal to the difference in energies of the outer shell and inner shell orbitals. For example, the photon emitted when neon loses an inner shell electron has an energy of 849 cv. The difference in energy between the 2*p* and 1*s* orbitals of neon is 869 ev. Photons with energies in this range occur in the X-ray region of the spectrum. It is apparent from the variation in the 1*s* orbital energies shown in Fig that the energies and hence the frequencies of the X-ray photons will increase as the nuclear charge is increased. It was from a study of the X-ray photons emitted by the elements that Moseley was first able to determine the nuclear charges (the atomic numbers) of the elements.

Some Chemical Implications

A detailed study of the chemical implications of the orbital theory of electronic structure must await our discussion of the chemical bond. However, we can at this point correlate the gross chemical behaviour of the elements with the general results of the orbital theory.

The effective nuclear charge is a minimum for the group I elements in any given row of the periodic table. Therefore, it requires less energy to remove an outer electron from one of these elements than from any other element in the periodic table. The strong reducing ability of these elements is readily accounted for. The variation in the relative reducing power of the elements across a given period or within a given group will be determined by the variation in the effective nuclear charge. The ability of the elements in a given row of the periodic table to act as reducing agents should undergo a continuous decrease from group I to group VII, since the effective nuclear charge increases across a given row. Similarly, the reducing ability should increase down a given column (group) in the table since the effective nuclear charge decreases as the principal quantum number is increased. Anticipating the fact that electrons can be transferred from one atom (the reducing agent) to another (the oxidizing agent) during a chemical reaction, we expect the elements to the left of the periodic table to exhibit a strong tendency to form positively charged ions.

The ability of the elements to act as oxidizing agents should parallel directly the variations in the effective nuclear charge. Thus the oxidizing ability should increase across a given row (from group I to group VII) and decrease down a given family. These trends are, of course, just the opposite of those noted for the reducing ability. We can also relate the chemical terms "reducing ability" and "oxidizing ability" to the experimentally determined energy quantities, "ionization potential" and "electron affinity." The reducing ability should vary inversely with the ionization potential, and the oxidizing ability should vary directly with the electron affinity. The elements in groups VI and VII should exhibit a strong tendency for accepting electrons in chemical reactions to form negatively charged ions. Francium, which possesses a single outer electron in the 7*s* orbital, should be the strongest chemical reducing agent and fluorine, with an orbital vacancy in the 2*p* subshell, should be the strongest oxidizing agent.

A great deal of chemistry can now be directly related to the electronic structure of the elements. For example, the reaction

$$Cl_2 + 2Br^- \rightarrow 2Cl^- + Br^2$$

is explained chemically by stating that Cl_2 is a stronger oxidizing agent than Br_2. The electronic interpretation is that the orbital vacancy in *Cl* is in a 3*p* orbital and closer to the nucleus than the 4*p* orbital vacancy in Br. Thus the effective nuclear charge which attracts the extra electron is larger for the *Cl* atom than for the Br atom. We could of course interpret this same reaction by stating that the Br^- ion is a stronger reducing agent than is the Cl^- ion. In other words the extra electron in the Br^- ion is less tightly held than is the extra electron in the Cl^- ion. The explanation in terms of the relative effective nuclear charges is the same as that given above.

The decrease in the effective nuclear charge down the halogen family of elements leads to some interesting differences in their chemistry. For example, hydrogen chloride may be prepared from sodium chloride and sulphuric acid:

$$2Cl^- + 2H^+ + SO_4^{-2} \rightarrow 2HCl + So_4^{-2}$$

However, the same method cannot be employed in the preparation of hydrogen bromide or hydrogen iodide. In the preparation of hydrogen bromide from sodium bromide,

$$2Br^- + 2H^+ + SO_4^{-2} \rightarrow SO_4^{-2} + 2HBr$$

some of the HBr reacts further,

$$2HBr + 2H^+ + So_4^{-2} \rightarrow 2H_2O + So_2 + Br_2$$

and the *HBr* is thus contaminated. In preparation of hydrogen iodide a further reaction again occurs:

$$8HI + 2H^+ + So_4^{-2} \rightarrow H_2S + 4I_2 + 4H_2O$$

Reactions are clearly redox reactions in which the halide ions reduce the sulphur in the SO_4^{-2} anion to a lower oxidation state. Since Cl has the highest effective nuclear charge, the Cl^- ion should be the weakest reducing agent of the three halide ions. Indeed, the Cl^- ion is not a strong enough reducing agent to change the oxidation state of S in SO_4^{-2}. The Br^- ion possesses an intermediate value for the effective nuclear charge and thus it is a stronger reducing agent than the Cl^- ion. The Br^- ion reduces the oxidation number of sulphur from (+6) to (+4). Since the I^- ion binds the extra electron least of all (the electron is in an $n = 5$ orbital and the effective nuclear charge of iodine is the smallest of the three), it should be the strongest reducing agent of the three halide ions. The I^- ion in fact reduces the sulphur from (+6) to (–2).

A word about oxidation numbers and electron density distributions is appropriate at this point. An oxidation number does not, in general, represent the formal charge present on a species. Thus S is not S^{+6} in the SO^{-2} ion, nor is it S^{-2} in the H_2S molecule. However, the average electron density in the direct vicinity of the sulphur atom does increase on passing from SO_4^{-2} to H_2S. From their relative positions in the periodic table it is clear that oxygen will have a greater affinity for electrons than sulphur. Thus when sulphur is chemically bonded to oxygen the electron density in the vicinity of the sulphur atom is decreased over what it was in the free atom and increased in the region of each oxygen atom. Again it is clear from the relative positions of H and S in the periodic table that sulphur has a greater affinity for electrons than does hydrogen. Thus in the molecule H_2S, the electron density in the vicinity of the sulphur atom is increased over that found in the free atom. In changing the immediate chemical environment of the sulphur atom from that of four oxygen atoms to two hydrogen atoms, the electron density (i.e., the average number of electrons) in the vicinity of the sulphur atom has increased. The assignment of actual oxidation numbers is simply a bookkeeping device to keep track of the number of electrons, but the sign of the oxidation number does indicate the direction of the flow of electron density. Thus sulphur has a positive oxidation number when combined with oxygen (the sulphur atom has lost electron density) and a negative one when combined with hydrogen (the electron density around sulphur is now greater than in the sulphur atom).

The above are only a few examples of how a knowledge of the electronic structure of atoms may be used to understand and correlate a large amount of chemical information. It should be remembered, however, that chemistry is a study of very complex interactions and the few simple concepts advanced here cannot begin to account for the incredible variety of phenomena actually observed. Our discussion has been based solely on energy, and energy alone never determines completely the course of a reaction on a macroscopic level, i.e., when many molecules undergo the reaction. There are statistical factors, determined by the changes in the number of molecules and in the molecular dimensions, which must also be considered. Even so, the energy effect can often be overriding.

In the long form of the periodic table, families are labelled by both a number and by the letter *A* or *B*. Thus there is a *IA* family and a *IB* family. It will be noted that the elements in a *B* family all occur in the series of transition elements in which the *d* orbitals are being filled. In the *A* families, however, the *d* orbitals are either absent or are present as closed inner shells. For example, consider the electronic configurations of *K* (*IA*) and *Cu* (*IB*):

$$K[Ar]4s^1$$

$$Cu[Ar]3d^{10}4s^1$$

Note that the most stable configuration for *Cu* is not [*Ar*] $3d^9 4s^2$ as expected. By transferring one of the 4*s* electrons to the 3*d* vacancy, the *d* subshell is filled and the electronic energy is lowered. The electron density distribution of the *Cu* atom is therefore a spherical one. Both *K* and *Cu* have one outer electron with a spherical charge distribution. They should have some properties in common, such as a tendency to lose one electron and form a positive ion. For this reason both families are labelled I. However, the shell underlying the outer electron in the *K* atom possesses a rare gas configuration, while in the *Cu* atom it is a set of filled *d* orbitals. This difference in electronic structure is sufficient to cause considerable differences in their chemistry, hence the further labels *A* and *B*.

A rare gas configuration is always one of great stability, particularly when it occurs in a positive ion. The species K^{+2} is never observed in solution chemistry, and could be produced in the gas phase only by an expenditure of energy far in excess of that observed in ordinary chemical reactions. The Cu^{+} ion, on the other hand, very readily loses a second electron to form the Cu^{+2} ion. Indeed, Cu^{+2} is the more common ionic form of copper. Thus the d^{10} closed shell structure is more easily broken than a rare gas configuration, giving to *Cu* a variable valency of one or two.

Chapter 4

The Chemical Bond

With our knowledge of the electronic structure of atoms we are now in a position to understand the existence of molecules. Clearly, the force which binds the atoms together to form a molecule will, as in the atomic case, be the electrostatic force of attraction between the nuclei and electrons. In a molecule, however, we encounter a force of repulsion between the nuclei in addition to that between the electrons. To account for the existence of molecules we must account for the predominance of the attractive interactions. We shall give general arguments to show that this is so, first in terms of the energy of a molecule, relative to the energies of the constituent atoms, and secondly, in terms of the forces acting on the nuclei in a molecule.

In order to determine what attractive and repulsive interactions are possible in a molecule, consider an instantaneous configuration of the nuclei and electrons in a hydrogen molecule. When the two atoms are initially far apart (the distance R is very large) the only potential interactions are the attraction of nucleus A for electron number (1) and the attraction of nucleus B for electron number (2). When R is comparable to the diameter of an atom (A and B are close enough to form a molecule) then new interactions appear. Nucleus A will now attract electron (2) as well as (1) and similarly nucleus B will attract electron (1) as well as (2). These interactions are indicated by the four solid lines in Fig. connecting pairs of particles which attract one another.

The number of attractive interactions has been doubled from what it was when the atoms were far apart. However, the reduction in R introduces two repulsive interactions as well, indicated by the dashed lines joining charges of like sign in Fig. The two electrons now repel one another as do the two nuclei. If the two atoms are to remain together to form a molecule, the attractive interactions must exceed the repulsive ones. It is clear from Fig that the new attractive interactions, nucleus A attracting electron (2) and nucleus B attracting electron (1), will be large only if there is a high probability of both electrons being found in the region between the nuclei. When in this region, both electrons are strongly attracted by both nuclei, rather than by just one nucleus as is the case when the atoms are far apart.

When the average potential energy is calculated by quantum mechanics, the attractive interactions are found to predominate over the repulsive ones because quantum mechanics does indeed predict a high probability for each electron being in the region between the nuclei. This general consideration of the energy demonstrates that electron density must be concentrated between the nuclei if a stable molecule is to be formed, for only in this way can the attractive interaction be maximized. We can be much more specific in our analysis of this problem if we discuss a molecule from the point of view of the forces acting on the nuclei. However, we must first state some general conclusions of quantum mechanics regarding molecular systems.

In the atomic case we could fix the position of the nucleus in space and consider only the motion of the electrons relative to the nucleus. In molecules, the nuclei may also change positions relative to one another. This complication can, however, be removed. The nuclei are very massive compared to the electrons and their average velocities are consequently much smaller than those possessed by the electrons. In a classical picture of the molecule we would see a slow, lumbering motion of the nuclei accompanied by a very rapid motion of the electrons. The physical implication of this large disparity in the two sets of velocities is that the electrons can immediately adjust to any change in the position of the nuclei. The positions of the nuclei determine the potential field in which the electrons move. However, as the nuclei change their positions and hence the potential field, the electrons can immediately adjust to the new positions. *Thus the motion of the electrons is determined by where the nuclei are but not by how fast the nuclei are moving.*

We may, because of this fact, discuss the motions of the electrons and of the nuclei separately.

For a given distance between the nuclei we obtain the energy, the wave function and the electron density distribution of the electrons, the nuclei being held in fixed positions. Then the distance between the nuclei is changed to a new value, and the calculation of the energy, wave function and electron density distribution of the electrons is performed again. This process, repeated for every possible internuclear distance, allows us to determine how the energy of the electrons changes as the distance between the nuclei is changed. More important for our present discussion, we may concern ourselves only with the motion of the electrons and hold the nuclei stationary at some particular value for the internuclear distance R.

The energy of the electrons in a molecule is quantized, as it is in atoms. When the nuclei are held stationary at some fixed value of R, there are a number of allowed energy levels for the electrons. There are, however, no simple expressions for the energy levels of a molecule in terms of a set of quantum numbers such as we found for the hydrogen atom. In any event we shall be concerned here only with the first or lowest of the energy levels for a molecule. As in the case of atoms, there is a wave function which governs the motion of all the electrons for each of the allowed energy levels. Each wave function again determines the manner in which the electronic charge is distributed in three-dimensional space.

The electron density distribution for a molecule is best illustrated by means of a contour map, of the kind introduced earlier in the discussion of the hydrogen atom. Figure shows a contour map of the charge distribution for the lowest, or most stable state of the hydrogen molecule. Imagine a hydrogen molecule to be cut in half by a plane which contains the nuclei.The amount of electronic charge at every point in space is determined, and all points having the same value for the electron density in the plane are joined by a line, a contour line. Also shown is a profile of the contour map along the internuclear axis.

The electron density contours of highest value are in the region of each nucleus. Thus the negative charge is concentrated in the region of the nuclei in a molecule as well as in an atom. The next highest concentration of negative charge is found in the region between the nuclei. It is the negative charge in this region which is strongly attracted by both nuclei and which results in the attractive interactions exceeding the repulsive ones in the formation of the molecule from the atoms. Most of the density contours envelope both nuclei. The density distributions of the two atoms have been merged together in the formation of the molecule.

The same contour map would be obtained for any plane through the nuclei. Therefore, in three-dimensional space the hydrogen molecule would appear to be an ellipsoidal distribution of negative charge. Most of the electronic charge is concentrated along the internuclear axis and becomes progressively more diffuse at large distances from the centre of the molecule. Recall that the addition of all the charge in every small volume element of space equals the total number of electrons which in the case of the hydrogen molecule is two. The volume of space enclosed by the outer contour in Fig. contains over 99 per cent of the total electronic charge of the hydrogen molecule.

An Electrostatic Interpretation of the Chemical Bond

In the light of the above discussion of a molecular electron density distribution, we may regard a molecule as two or more nuclei imbedded in a rigid three-dimensional distribution of negative charge. There is a theorem of quantum mechanics which allows us to make direct use of this picture of a molecule. This theorem states that the force acting on a nucleus in a molecule may be determined by the methods of classical electrostatics. The nuclei in a molecule repel one another, since they are of like charge. This force of repulsion, if unbalanced, would push the nuclei apart and the molecule would separate into atoms. In a stable molecule, however, the nuclear force of repulsion is balanced by an attractive force exerted by the negatively-charged electron density distribution. The usefulness of this approach lies in the fact that we may account for and discuss the stability of molecules in terms of the classical concept of a balance between the electrostatic forces of attraction and repulsion. We can illustrate this method and arrive at some results of a general nature by considering in detail the forces acting on the nuclei in the hydrogen molecule.

The charge on a hydrogen nucleus is $+e$ and the force of repulsion acting on either nucleus is

$$F_n = +e^2 / R^2$$

where R is the internuclear distance. This force obviously acts to push the two nuclei apart.

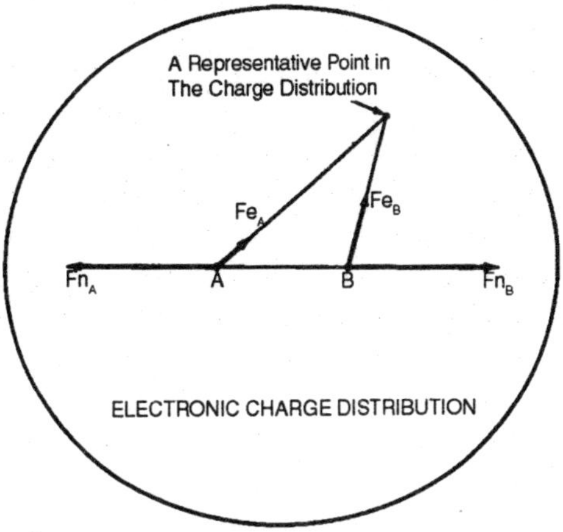

Fig. The Forces Acting on The Nuclei in H_2.

The attractive force which balances this force of repulsion and draws the nuclei together is exerted by the negatively-charged electron density distribution. The density distribution is treated as a *rigid* distribution of negative charge in space. Each small element of this charge distribution exerts a force on the nuclei, illustrated in Fig. for one such small charge point. The forces it exerts on the nuclei are labelled F_{eA} and F_{eB}. The total amount of negative charge in the electron density distribution must correspond to some integral number of electrons. However, the amount of negative charge in each small region of space will in general correspond to some fraction of one electronic charge.

The electronic force of attraction F_{eA} or F_{eB} may be equated to two components, one along the bond, and one perpendicular to it. The density distribution is symmetric with respect to the internuclear axis, i.e., for every charge point above the axis there must, by symmetry, be another point of equal charge at the corresponding place beneath the internuclear axis. The symmetrically related charge point will exert the same force along the bond, but the component perpendicular to the bond will be in the opposite direction. Thus the perpendicular forces of attraction exerted on the nuclei are zero and we may confine our attention to the components of the attractive force along the bond.

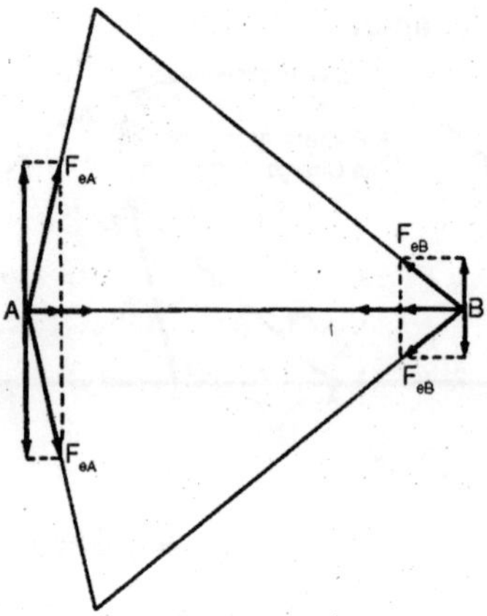

Fig. The Two Components of Force along The Bond add Together while The two Perpendicular Components cancel at both *A* and *B*.

It is obvious that all of the charge elements which are in the general region between the two nuclei will exert forces which draw the two nuclei together. The force exerted by the density in this region acts in opposition to the force of nuclear repulsion and binds the two nuclei together. It is also clear that a charge element in the region behind either nucleus will exert a force which tends to increase the distance between the nuclei.

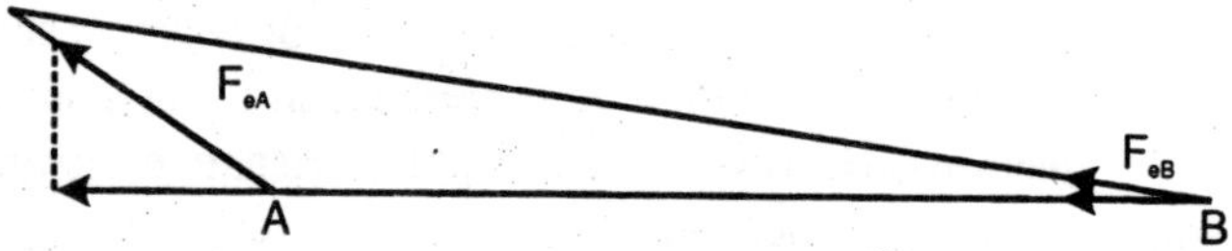

Fig. The Component of F_{eA} along The Bond is Greater than The Corresponding Component of F_{eB}.

Since the charge element is closer to nucleus *A* than it is to nucleus *B*, the component of the force on *A* along the bond will be greater than the component of the force on *B* along the bond. Thus the effect of density in this region will be to separate the molecule into atoms.

There must also be a line on which the density exerts the same force on both nuclei and thus neither increases nor decreases R because the charge density in one region draws the nuclei together and in another draws them apart. The charge element shown in Fig exerts the same force along the bond on both A and B even though it is closer to B than it is to A. Although the total force F_{eB} is much larger than F_{eA}, F_{eB} is directed almost perpendicular to the bond axis and thus its component along the bond is quite small and equal to the component of F_{eA} along the bond. Charge density on either of the two curves shown in Fig. exerts equal forces on both of the nuclei along the bond, and such charge density will not tend to increase or decrease the distance between the nuclei. Thus these two curves (surfaces in three dimensions) divide the space in a molecule into a binding region and an antibinding region. Any charge density between the two boundary curves, in the *binding region*, draws the two nuclei together while any charge density in the hatched region behind either curve, the *antibinding region*, exerts unequal forces on the nuclei and separates the molecule into atoms.

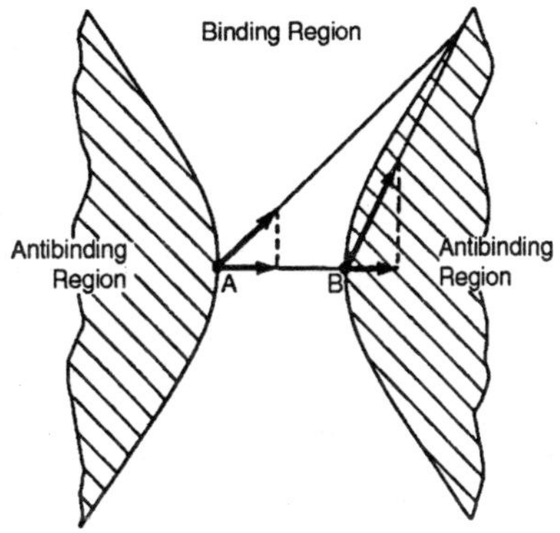

Fig. The Boundary Curves which separate the Binding from The Antibinding regions in a Homonuclear Diatomic Molecule.

A chemical bond is thus the result of the accumulation of negative charge density in the region between the nuclei to an extent sufficient to balance the nuclear forces of repulsion. This corresponds to a state of electrostatic equilibrium as the net force acting on each nucleus is zero for this one particular value of the internuclear distance. If the distance between the nuclei is increased from the equilibrium value, the nuclear force of repulsion is decreased. At the same time the force of attraction exerted by the electron density distribution is increased as the binding region is increased in size. Thus when R is increased from its equilibrium value there are net forces of attraction acting on the nuclei which pull the two nuclei together again. A definite force would have to be applied to overcome the force of attraction exerted by the electron density distribution and separate the molecule into atoms. Similarly, if the value of R is decreased from its equilibrium value, the force of nuclear repulsion is increased over its equilibrium value. At the same time, the attractive force exerted by the electron density is decreased, because the binding region is decreased in size. In this case there will be a net force of repulsion pushing the two nuclei apart and back to their equilibrium separation. There is thus one value of R for which the forces on the nuclei are zero and the whole molecule is in a state of electrostatic equilibrium.

The division of the space around a molecule into a binding and an antibinding region shows where charge density must be concentrated in order to obtain a stable chemical bond. The next question which must be answered is, "How much charge must be placed in the binding region to achieve electrostatic equilibrium?" For example, we might consider the possibility of forming a molecule by bringing together two atoms, each with its own atomic distribution of charge, and simply allow the two atomic charge distributions to overlap without deforming in any way. This would result in the accumulation of approximately twice as much charge density in the binding region as in either of the antibinding regions behind the nuclei. Would this doubling of the charge density in the region between the nuclei be sufficient to balance the nuclear forces of repulsion? Let us answer this question for the simple case of two hydrogen atoms forming molecular hydrogen, but again the result will be general.

The most stable state of the hydrogen molecule is obtained when two hydrogen atoms, each in its most stable atomic state, approach one another. The ground state of a hydrogen atom is obtained when the electron is in the $1s$ orbital. The density distribution around each hydrogren nucleus is the spherical one which we discussed previously in some detail. We shall first calculate the force on one of the hydrogen nuclei resulting when the two atoms are very far apart. The situation is represented in Fig where each atomic charge distribution is represented by a single outer circular contour. This contour is to define a sphere which in three-dimensions contains essentially all of the electronic charge of each atom.

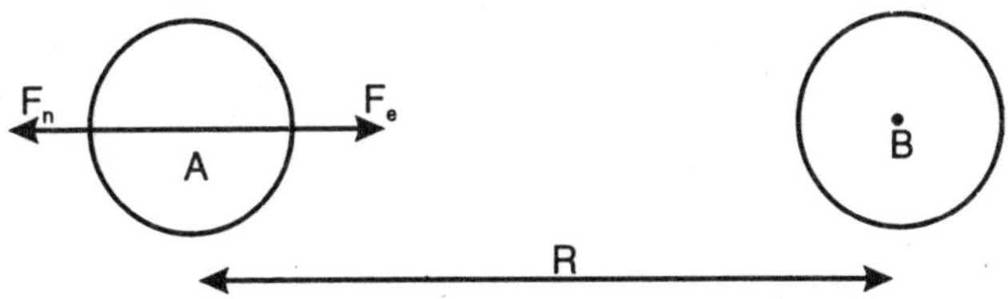

Fig. The Forces Acting on Nucleus *A* at a large Internucleus Distances, *R*.

Consider the forces exerted on nucleus *A*. The force of nuclear repulsion is just

$$F_n = (+e)(+e)/R^2 = e^2/R^2$$

The atomic charge density centred on nucleus *A* exerts no net force on this nucleus as it pulls equally in every direction because of its spherical symmetry. There is, however, a net force of attraction due to the single electronic charge dispersed in the atomic distribution of *B*. A theorem of classical electrostatics states that the force exerted by a spherical charge distribution on a point charge lying outside of the charge distribution is equal to the force which would be obtained if all the charge in the distribution were concentrated at its centre. Nucleus *A* is a point charge which lies outside of the spherical charge distribution centred on B. Thus the force exerted on nucleus A by this charge distribution is just

$$F_e = \frac{(-e)(+e)}{R^2} = -e^2/R^2$$

as the total amount of charge contained in the distribution is that of one electron. The total force acting on nucleus *A* is

$$F_n + F_e = e^2/R^2 - e^2/R^2 = 0$$

A zero force is the expected answer when the two atoms are very far apart.

Can we again balance the forces for a value of R which is of the order of magnitude of an atomic diameter, i.e., typical of the values of R found in molecules? At this value of R, each nucleus will have penetrated the charge density surrounding the other nucleus. Recall that in this calculation we insist upon the atomic charge densities remaining spherical and our molecular charge density is obtained by allowing the two rigid atomic charge distributions to overlap one another.

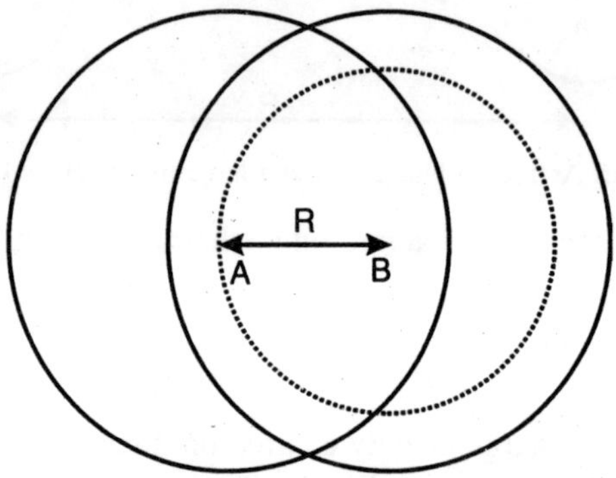

Fig. The forces exterted on the nucleus A for the overlap of rigid atomic charge distributions. Only the charge density on B which is contained in the sphere of radius R exerted a force on nucleus A.

The force of nuclear repulsion in this case is still given by

$$F_n = e^2 / R^2$$

where the value of R is much less than in the previous calculation. Since the charge distribution on A is still spherical in shape, it exerts no net force on nucleus A. The force exerted on nucleus A by the charge density on B can again be calculated by the theorem referred to previously. However, nucleus A no longer lies outside of all the charge density on B. The value of R is significantly less than the radius of the charge distribution on B. All the charge density on B which lies within the sphere defined by the bond length R again exerts a force on nucleus A, equal to that obtained if all this density were situated at the B nucleus. The theorem referred to previously shows that the density on *B* which lies outside of this sphere defined by *R* exerts no net force on nucleus *A*.

Since *R* is less than the diameter of the charge distribution, the amount of negative charge contained in a sphere of radius *R* will be less than that of one electron. The observed value of *R* for the hydrogen molecule is 1.4 au and reference to the data given previously for the 1*s* orbital density for the hydrogen atom shows that a sphere of radius 1.4 would contain approximately one half of an electronic charge. The electrostatic force of attraction exerted on nucleus A is, therefore,

$$F_e = -0.5e^2 / R^2$$

The net force on nucleus *A* is

$$\begin{aligned} F &= F_n + F_e \\ &= +e^2 / R^2 - 0.5 / R^2 \\ &= 0.5e^2 / R^2 \end{aligned}$$

There is a net force of repulsion exerted on nucleus *A* under these conditions. If *R* were decreased still further, nucleus *A* would penetrate the charge density around *B* to an even greater extent and "see" even more of the nuclear charge on *B*. The force on the nuclei will thus be repulsive for all finite values of *R*.

This is an important result as it shows that the density distribution in a molecule cannot be considered as the simple sum of the two atomic charge densities. The overlap of rigid atomic densities does not place sufficient charge density in the binding region to overcome the nuclear force of repulsion. We conclude that the original atomic charge distributions must be distorted in the formation of a molecule, and the distortion is such that charge density is concentrated in the binding region between the nuclei. A quantum mechanical calculation predicts this very result. The calculation shows that there is a continuous distortion of the original atomic density distributions, a distortion which increases as the internuclear distance decreases. This is illustrated in Fig. for the approach of two hydrogen atoms to form the hydrogen molecule.

The changes in the original atomic density distributions caused by the formation of the chemical bond may be isolated and studied directly by the construction of a *density difference distribution.* Such a distribution is obtained by subtracting the density obtained from the overlap of the undistorted atomic densities separated by a distance *R*, from the molecular charge distribution evaluated at the same value of *R*. Wherever this density difference is positive in value it means that the electron density in the molecule is greater than that obtained from the simple overlap of the original atomic densities. Where the density difference is negative, it means that there is less density at this point in space in the molecule than in the distribution obtained from the overlap of the original atomic distributions. Such a density difference map thus provides a detailed picture of the net reorganization of the charge density of the separated atoms accompanying the formation of a molecule.

We have just proven that the density distribution resulting from the overlap of the undistorted atomic densities does not place sufficient charge density in the binding region to balance the forces of nuclear repulsion. The regions of charge increase in the density difference maps are, therefore, the regions to which charge is transferred relative to the separated atoms to obtain a state of electrostatic equilibrium and hence a chemical bond. From this point of view a density difference map provides us with a picture of the "bond density."

A set of density difference or bond density maps for the approach of two hydrogen atoms to form the hydrogen molecule. At very large separations, for example at 8 au, the density distribution on each atom is polarized in the direction of the approaching atom. Charge density has been transferred from the antibinding region behind each nucleus to the binding region immediately in front of each nucleus. Thus even at large separations the atomic density distributions are no longer spherical. We noted in our discussion of the approach of two rigid hydrogen atoms that a spherical charge distribution does not exert a net force on the nucleus on which it is centred. Each polarized atomic charge distribution does, however, exert an attractive force on its nucleus. The polarized densities place more charge on the binding side of each nucleus than on the antibinding side. These long-range attractive forces, called van der Waals' or dispersion forces, could be aptly described as a "bootstrap effect" as each nucleus is pulled by its own charge density. All pairs of neutral molecules undergo this type of polarization as a result of the long-range interactions between them, and there are attractive forces operative between all pairs of molecules out to very large distances. Although the long-range polarizations and the resulting forces of attraction are very weak, they are of extreme importance. They are commonly referred to as van der Waals forces and are solely responsible for the binding observed in certain kinds of solids, solid helium for example.

At 6.0 au the density increase in the binding region is common to both nuclei, and for distances less than 6.0 au the system can no longer be described as two polarized hydrogen atoms. The distortions of the original densities caused by the transfer of charge to the binding region is so great that the individual character of the atomic densities is no longer discernible. The *magnitude* of the attractive force (which is negative in sign) exerted on the nuclei by this accumulation of charge density in the binding region increases rapidly for distances less than 4.5 au.

The attractive force reaches a maximum at 2.1 au. The density difference diagrams indicate that for distances as small as 2.0 au, the density increase is confined to the region between the nuclei. For separations smaller than 2.0 au an increasing amount of charge density is transferred to the anti-binding regions behind each nucleus. Because of this, the attractive force on the nuclei decreases rapidly with a further decrease in R until at R = 1.4 au, the net attractive force exerted by the charge density just balances the force of nuclear repulsion. A state of electrostatic equilibrium is reached, and a chemical bond is formed. A further decrease in R leads to a force of repulsion. More charge density is transferred to the antibinding regions, and the force exerted by this charge density, acting in concert with the increase in the force of nuclear repulsion, outweighs the attractive force exerted by the charge density in the binding region.

The same changes in density are depicted in Fig, which is a series of profiles along the internuclear axis of the density difference maps shown in Fig. The profile maps illustrate in a striking fashion the build-up of charge density in the region between the nuclei.

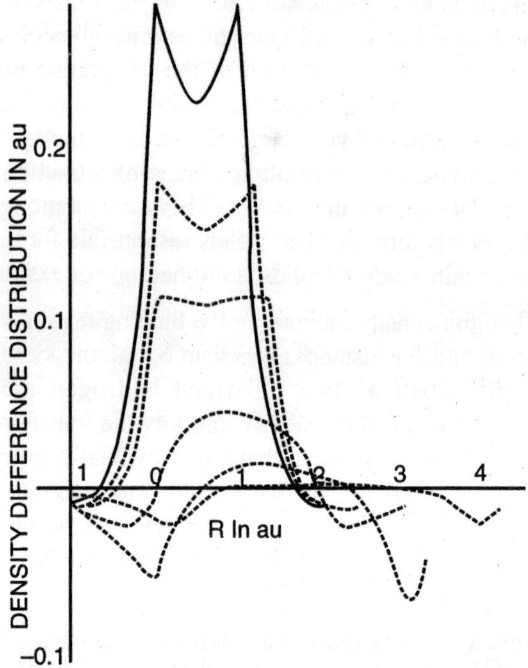

Fig. Profiles of The Density Difference along the Internuclear Axis for H_2

The formation of any chemical bond is qualitatively similar to the changes in the charge distribution and in the forces exerted on the nuclei as found for the hydrogen molecule. We must now inquire into the conditions which determine whether or not sufficient charge density can be accumulated in the binding region to yield a stable molecule. Since not all atoms form chemical bonds, clearly such conditions must exist.

The Effect of the Pauli Principle on Chemical Binding

The Pauli exclusion principle plays as important a role in the understanding of the electronic structure of molecules as it does in the case of atoms. The end result of the Pauli principle is to limit the amount of electronic charge density that can be placed at any one point in space. For example, the Pauli principle prevents the 1*s* orbital in an atom from containing more than two electrons. Since the 1*s* orbital places most of its charge density in regions close to the nucleus, the Pauli principle, by limiting the occupation of the $1s$ orbital, limits the amount of density close to the nucleus. Any remaining electrons must be placed in orbitals which concentrate their charge density further from the nucleus.

In an earlier discussion we pointed out that the reason the electron doesn't fall onto the nucleus is because it must possess kinetic energy if Heisenberg's uncertainty principle is not to be violated. This is one reason why matter doesn't collapse. The Pauli principle is equally important in this regard. The electron density of the outer electrons in an atom cannot collapse and move closer to the nucleus since it can do so only if the electrons occupy an orbital with a lower n value. If, however, the inner orbital contains two electrons, then the Pauli principle states that the collapse cannot occur. We must be careful in our interpretation of this aspect of the Pauli principle. The density from a $2s$ orbital has a small but finite probability of being found well within the density of the $1s$ orbital. Do not interpret the Pauli principle as implying that the density from an occupied orbital has a clearly defined and distinct region in real space all to its own. This is not the case. The operation of the Pauli principle is more subtle than this. In some simple cases, such as the ones we wish to discuss below, the limiting effect of the Pauli principle on the density distribution can, however, be calculated and pictured in a very direct manner.

The Pauli principle demands that when two electrons are placed in the same orbital their spins must be paired. What restriction is placed on the spins of the electrons during the formation of a molecule, when two orbitals, each on a different atom, overlap one another? For example, consider the approach of two hydrogen atoms to form a hydrogen molecule. Consider atom A to have the configuration $1s^1(\uparrow)$and atom B the configuration $1s^1(\downarrow)$. Even when the atoms approach very close to one another the Pauli principle would be satisfied as the spins of the two electrons are opposed. This is the situation we have tacitly assumed in our previous discussion of the hydrogen molecule. However, what would occur if two hydrogen atoms approached one another and both had the same configuration and spin, say $1s^1(\uparrow)$? When two atoms are relatively close together the electrons become indistinguishable. It is no longer possible to say which electron is associated with which atom as both electrons move in the vicinity of both nuclei. Indeed this is the effect which gives rise to the chemical bond. In so far as we can still regard the region around each atom to be governed by its own atomic orbital, distorted as it may be, two electrons with the same spin will not be able to concentrate their density in the binding region. This region is common to the orbitals on both atoms, and since the electrons possess the same spin they cannot both be there simultaneously. In the region of greatest overlap of the orbitals, the binding region, the presence of one electron will tend to exclude the presence of the other if their spins are parallel. Instead of density accumulating in the binding region as two atoms approach, electron density is removed from this region and placed in the antibinding region behind each nucleus where the overlap of the orbitals is much smaller. Thus the approach of two hydrogen atoms with parallel spins does not result in the formation of a stable molecule. This repulsive state of the hydrogen molecule, in which both electrons have the same spin and atomic orbital quantum numbers, can be detected spectroscopically.

We can now give the general requirements for the formation of a chemical bond. Electron density must be accumulated in the region between the nuclei to an extent greater than that obtained by allowing the original atomic density distributions to overlap. In general, the increase in charge density necessary to balance the nuclear force of repulsion requires the presence of two electrons.

In the atomic orbital approximation we picture the bond as resulting from the overlap of two distorted atomic orbitals, one centred on each nucleus. When the orbitals overlap, both electrons may move in the field of either nuclear charge as the electrons may now exchange orbitals. Finally, the pair of electrons must possess opposed spins. When their spins are parallel, the charge density from each electron is accumulated in the antibinding rather than in the binding region.

We shall now apply these principles to a number of examples and in doing so obtain a quantum mechanical definition of valency.

The Quantum Mechanical Explanation of Valency

Helium atoms in their ground state do not form a stable diatomic molecule. In fact, helium does not combine with any neutral atom. Its valency, that is, its ability to form chemical bonds with other atoms, is zero. The electronic configuration of the helium atom is $1s^2(\uparrow\downarrow)$, a closed shell configuration. When two helium atoms are in contact, each electron on one atom encounters an electron on the other atom with a parallel spin. Because of the Pauli principle, neither electron on either atom can concentrate its density in the region they have in common, the region between the nuclei. Instead, the density is transferred to the antibinding regions behind each nucleus where the overlap of the two atomic density distributions is least. This is the same effect noted earlier for the approach of two hydrogen atoms with parallel spins.

Comparison of a series of density difference maps for the approach of two helium atoms with those given previously for H_2 reveals that one set is the opposite of the other. The regions of charge build-up and charge depletion are reversed in the two cases. The density difference diagrams are obtained by subtracting the distribution obtained by the overlap of the atomic charge densities from the molecular charge distribution. The former distribution, it will be recalled, does not place sufficient charge density in the binding region to balance the force of nuclear repulsion. Thus it is clear from Fig that He_2 will be unstable because the molecular distribution places less charge density in the binding region than does the one obtained from the overlap of the atomic densities. The charge density in He_2 is transferred to the antibinding region where it exerts a force which, acting in the same direction as the nuclear force of repulsion, pulls the two nuclei apart. Repulsive forces will dominate in He_2 and no stable molecule is possible.

A comparison of the density difference profiles for He_2 and H_2 provides a striking contrast of the difference between the charge redistributions which result in the formation of unstable and stable molecules.

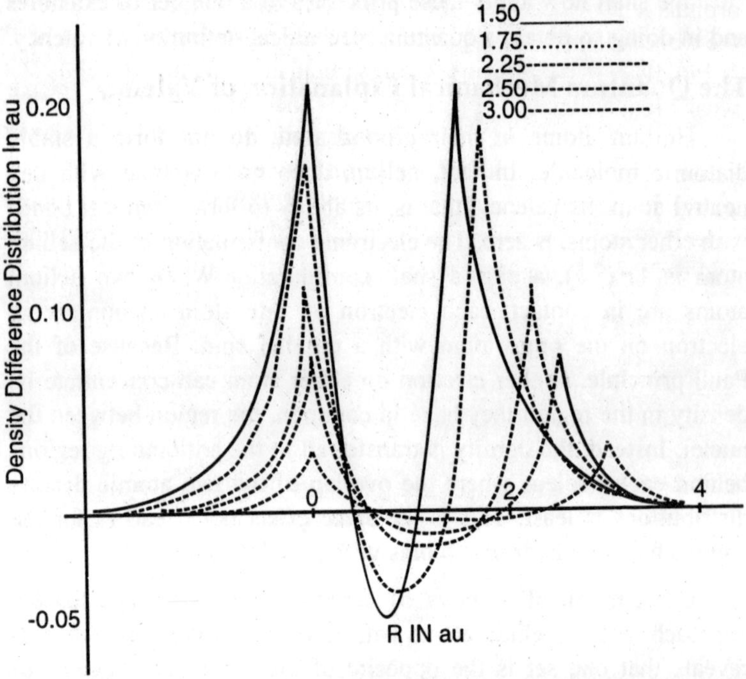

Fig. Profiles of The Density Difference Maps along The Internuclear Axis for The Approach of two He Atoms.

The force on a helium nucleus in He_2 as a function of the internuclear separation is repulsive for the range of *R* values indicated.

Unlike the force curve for H_2, there is no deep minimum in the curve which represents a range of *R* values for which the force is attractive. The force curve for He_2 does cross the *R* axis at approximately 6 au (this is not indicated in Fig. and becomes very slightly attractive for values of *R* greater than this value. This weak attractive force has its origin in the long-range mutual polarization of the atomic density distributions which was discussed in detail for the approach of two hydrogen atoms. For large internuclear separations, where there is no significant overlap of the atomic orbitals and hence no need to invoke the Pauli exclusion principle, the atomic charge distributions of two approaching helium atoms are polarized in the same way as are the charge distributions for two approaching hydrogen atoms, and the force is attractive. At smaller internuclear separations, however, where the overlap of the orbitals is significant and the Pauli exclusion principle is operative, the direction of the charge transfer in He_2 is reversed and the force is rapidly transformed into one of repulsion. Were it not for the weak long-range attractive forces - the van der Waals forces - gaseous helium could not be condensed into a liquid or a solid phase. As it is, the force of attraction between two helium atoms is so weak that at a temperature of only 4.2°K they have sufficient kinetic energy to overcome the forces of attraction between them and escape into the gas phase.

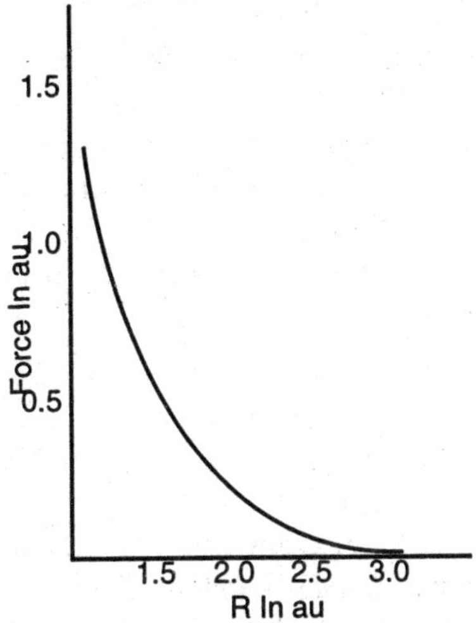

Fig. Force an a *He* Nucleus in He_2 as a Function of The Internuclear Separation.

If it was not necessary to satisfy the demands of the Pauli principle, electron density would accumulate in the binding region of He_2, even for small values of *R*, as this region is of lower potential energy than is the antibinding region. However, when each electron detects another of like spin (when the orbitals overlap) they cannot concentrate their charge density in the region they have in common, the binding region. That it is indeed the Pauli principle which prevents the formation of He_2 is evident from the fact that He_2^+, which possesses one less electron, *is stable*! When a helium atom approaches a helium ion, an orbital vacancy is present and the density from one pair of electrons (those with opposed spins) can be concentrated in the binding region.

All the rare gas atoms possess a closed shell structure and this accounts for their inertness in chemical reactions. No homonuclear diatomic molecules are found in this group of elements; all occur naturally in the atomic state. Compounds of *Kr* and *Xe* have been formed with fluorine, for the same reason that the formation of He_2^+ is possible. Fluorine has a very high electron affinity and a single vacancy in its outer quantum shell. Thus one of the electrons in the closed shell structure of *Xe* can be pulled into the orbital vacancy of the fluorine atom and density concentrated in the region between the nuclei.

Only an atom with a very high affinity for electrons will bond with a rare gas atom. The only species found with sufficient electron affinity to bind a helium atom (which holds its electrons the most tightly of all atoms) is a He^+ ion. If the helium atom has the highest ionization potential of all the elements, then the singly-charged He^+ ion must possess the highest electron affinity of all the neutral or singly-charged atoms.

Let us now attempt to explain the variation in the valency exhibited by the elements in the second row of the periodic table. The hydrides of these elements are LiH, BeH_2, BH_3, CH_4, NH_3, OH_2 and FH. The valency of the hydrogen atom is unity as it possesses one unpaired electron and one orbital vacancy. It can form one electron pair bond. Therefore, the valencies exhibited in the above hydrides must be 1, 2, 3, 4, 3, 2, 1, as this is the number of hydrogens bound in each case.

Fluorine. The electron configuration of F is $1s^2 2s^2 2p^5(\uparrow)$. Only one of the electrons in the $2p$ orbitals is unpaired. The $2p$ atomic orbital with the vacancy may overlap with the $1s$ atomic orbital of hydrogen, and if the spin of the electron on H is paired with the spin of the electron on F, all the requirements for the formation of a stable chemical bond will be met. The valency of F will be one as it possesses one unpaired electron and can form one electron pair bond.

Oxygen. The electronic configuration of oxygen is $1s^2 2s^2 2p^4(\uparrow\uparrow)$. Oxygen has two unpaired electrons, both of which may pair up with an electron on a hydrogen atom. The valency of oxygen should be two as is observed. It is obvious that all the requirements for a chemical bond can be met for every unpaired electron present in the outer or "valency" shell of an atom. *Thus valency may be defined as being equal to the number of unpaired electrons present in the atom.*

Nitrogen. The configuration of nitrogen is $1s^2 2s^2 2p^3(\uparrow\uparrow\uparrow)$, and its hydride should be NH_3 as is indeed the case.

Carbon. Since the most stable electron configuration of carbon is $1s^2 2s^2 2p^2(\uparrow\uparrow)$ we predict its valency to be two. The molecule CH_2 (called methylene) is indeed known. However, CH_2 is very reactive and its products are not stable until four chemical bonds are formed to carbon as in the case of CH_4. Four, not two, is the common valency for carbon. How can our theory account for this fact? The energy of a $2p$ orbital is not much greater than that of a $2s$ orbital.

Carbon in the promoted state possesses four unpaired electrons and can now combine with four hydrogen atoms. Every bond to a hydrogen atom releases a large amount of energy. The energy required to unpair the $2s$ electrons and promote one of them to a $2p$ orbital is more than compensated for by the fact that *two* new *C—H* bonds are obtained.

Boron. Boron has the electronic configuration $1s^2 2s^2 2p^1(\uparrow)$. Its valency should be one and BH is known to exist. However, again through the mechanism of promotion, the valency of boron can be increased to three:

B* $1s^2 2s^1(\uparrow) 2p^2(\uparrow\uparrow)$

We might wonder why, with a $2p$ orbital still vacant, one of the $1s$ electrons is not promoted and thus give boron a valency of five. This does not happen because of the large difference in energy between the $1s$ and $2p$ orbitals.

Beryllium. Beryllium has the configuration $1s^2 2s^2$ and should exhibit a valency of zero. The outer electron configuration of Be is similar to that of He, a closed shell of s electrons. Indeed, the molecule Be_2 exists only as a weakly bound van der Waals molecule. However, *Be* differs from He in that there are vacant orbitals available in its *valency* shell. The observed valency of two in the molecule BeH_2 can be explained by a promotion to the configuration $1s^2 2s^1(\uparrow)2p^1(\uparrow)$.

Lithium. Lithium, with the configuration $1s^2 2s^1(\uparrow)$, should exhibit only a valency of one.

Lewis Structures

The concept of an electron pair bond is not restricted to bonds with hydrogen. The only requirements are an unpaired electron on each atom (which is another way of saying there is an orbital vacancy on each atom) with their spins opposed. Thus two fluorine atoms may combine to form the fluorine molecule F_2 through the overlap of the singly-occupied $2p$ orbital on one atom with a similar orbital on the other. This will result in F_2 being described as *F—F* where the single line denotes that one pair of electrons forms the bond between the two atoms. Similarly, the three singly-occupied $2p$ orbitals on one nitrogen atom may overlap with those on another to form the N_2 molecule. Since three pairs of electrons are shared between the nuclei in this case, we represent the molecule by the symbol $N{\equiv}N$. The electrons in the valence shell of an atom which are not involved in the formation of a chemical bond (as they are already paired in an orbital on the atom) may also be indicated and the resulting symbols are called *Lewis structures*. Thus the three pairs of valency electrons on each F, $(2s^2 2p^4)$, not involved in the bonding are often indicated by dots. For example,

$$H \text{———} \ddot{\underset{\cdot\cdot}{F}}\!: \qquad :\!\ddot{\underset{\cdot\cdot}{F}} \text{———} \ddot{\underset{\cdot\cdot}{F}}\!: \qquad Li \text{———} \ddot{\underset{\cdot\cdot}{F}}\!:$$

(Lithium has only one outer electron and it is shared in the bond.) In compounds with nitrogen we may indicate the $2s$ pair of electrons:

$:NH_3$ and $:N \text{———} :N$

Recall that each line, since it denoted a bond in these diagrams, represents a pair of electrons shared between the two atoms joined by the line. If we add up the lines joined to each atom, multiply by two (to obtain the number of electrons) and add to this the number of dots which represents the remaining valence electrons, the number eight is obtained in many cases, particularly for the second-row elements (n = 2 valence orbitals). This so-called octet rule results from many elements having four outer orbitals ($nsnp_xnp_ynp_z$) which together may contain a total of eight electrons. Not all eight electrons belong to either atom in general as the electrons in a bond are shared (not necessarily equally as we shall see) between two atoms. Each bond contains two electrons with paired spins. Thus the orbital from one atom used to form the bond is, in a sense, filled as both spin possibilities are now accounted for.

The presence of an unshared pair of electrons in the valency shell of an atom can lead to the formation of another chemical bond. For example, the unshared pair of electrons in the 2s orbital on nitrogen in ammonia may attract and bind to the molecule another proton:

$$H_3N: + H^+ \rightarrow NH_4^+$$

A similar reaction occurs for the water molecule which possesses two unshared pairs of electrons:

$$H_2\ddot{\underset{\cdot\cdot}{O}} + H^+ \rightarrow H_3O:$$

We must modify our previous rule regarding the requirements for the formation of an electron pair bond. Rather than both orbitals being half-filled, an orbital on one of the atoms may be filled if the orbital on the other atom is completely vacant. Molecules possessing an unshared pair of electrons, which may be used to bond another atom, are called *Lewis bases*. Only elements in groups V, VI and VII will exhibit this property. The elements in groups I to IV do not possess unshared pairs. Instead, the chemistry of the elements in groups II and III is largely characterized by the orbital vacancies which they possess in their valency shell.

The compound boron trifluoride represents the pairing of the three valence electrons of boron with the unpaired electrons on three F atoms. The boron is considered to be in the promoted configuration $1s^2 2s^1(\uparrow) 2p^2(\uparrow\uparrow)$ and BF_3 is represented as

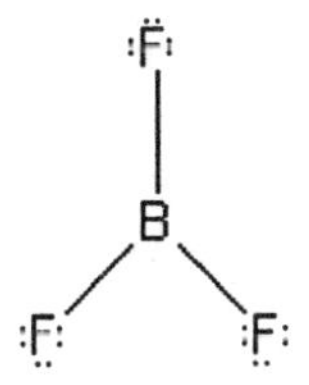

A $2p$ orbital on boron is vacant. It is not surprising to find that BF_3 may form another bond with a species which has an unshared pair of electrons, i.e., a Lewis base. For example,

$$BF_3 + :NH_3 \rightarrow H_3N—BF_3$$

Since BF_3 accepts the electron pair it is termed a *Lewis acid.* Further examples from group IIIA are

$$BF_3 + F^- \rightarrow BF_3^-$$

$$AlCl_3 + Cl^- \rightarrow AlCl_4^-$$

$$BH_3 + H^- \rightarrow BH_4^-$$

and from group IIA (which have two orbital vacancies):

$$BeCl_2 + 2Cl^- \rightarrow BeCl_4^{-2}$$

Molecular Geometry

The theory of valency which we have been developing is known as *valence bond theory.* One further feature of this theory is that it may be used to predict (or in some cases, rationalize) the observed geometries of molecules By the geometry of a molecule we mean the relative arrangement of the nuclei in three-dimensional space. For example, assuming the two *O—H* bonds in the water molecule to be similar and hence of the same length, the angle formed by the two *O—H* bonds (the *HOH* angle could conceivably posess any angle from 180° to some relatively small value. All we demand of our simple theory is that it correctly predict whether the water molecule is linear (bond angle = 180°) or bent (bond angle less than 180°) Or as another example, it should predict whether the ammonia molecule is planar (a) or pyramidal (b).

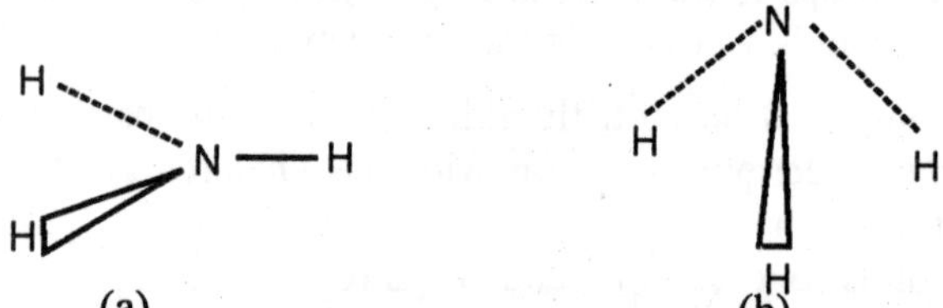

(a) (b)

The observed geometry of a molecule is that which makes the energy of the system a minimum. Thus those geometries will be favoured which (i) concentrate the largest amount of charge density in the binding region and thus give the strongest individual bonds, and (ii) keep the nuclei as far apart as possible (consistent with (i)), and hence reduce the nuclear replusions. Consider again the two possibilities for the water molecule It is clear that the linear form (a) will have a smaller energy of nuclear repulsion from the hydrogens than will the bent form (b).

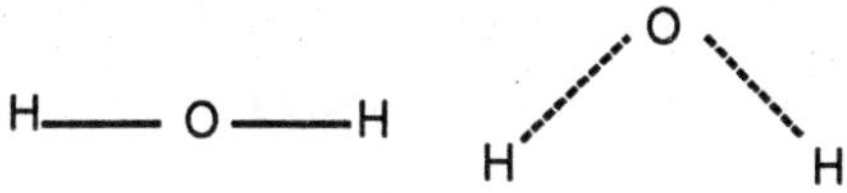

(a) (b)

If the amount of electron density which could be concentrated in the regions between the nuclei in each *O—H* bond (i.e., the strength of each *O—H* bond) was independent of the bond angle, then clearly the linear form of the water molecule would be the most stable. This would be the situation if all the atomic orbitals which describe the motions of the electrons were rigidly spherical and centred on the nuclei. But this is not the case. As was stressed earlier in our discussion of atomic orbitals, the motion of electrons possessing angular momentum because they occupy orbitals with $l \neq 0$ is concentrated along certain axes or planes in space. In particular the three *p* orbitals are a maximum along the three perpendicular axes in space. The valence bond theory of the water molecule describes the two *O—H* bonds as resulting from the overlap of the H 1*s* orbitals with the two half-filled 2*p* orbitals of the oxygen atom. Since the two 2*p* orbitals are at right angles to one another, valence bond theory predicts a bent geometry for the water molecule with a bond angle of 90°.

The overlap of the orbitals is shown schematically in Fig.

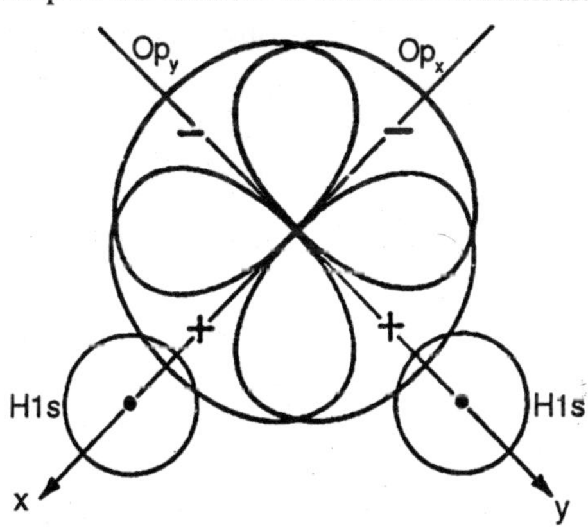

Fig. A pictorial representation of the overlap of $2p_x$ and $2p_y$ orbitals on the oxygen with the 1*s* orbitals of two H atoms.

The actual bond angle in the water molecule is 104.5°. The opening of the angle to a value greater than the predicted one of 90° can be accounted for in terms of a lessening of the repulsion between the hydrogen nuclei. The *assumption* we have made is that the maximum amount of electron density will be transferred to the binding region and hence yield the strongest possible bond when the hydrogen and oxygen nuclei lie on the axis which is defined by the direction of the $2p$ orbital. For a given internuclear separation, this will result in the maximum overlap of the orbitals. Because an orbital with $l \neq 0$ restricts the motion of the electron to certain preferred directions in space, bond angles and molecular geometry will be determined, *to a first rough approximation*, by the inter-orbital angles.

In the valence bond description of ammonia, each N—H bond results from the overlap of an H $1s$ orbital with a $2p$ orbital on N. All three $2p$ orbitals on N have a vacancy and thus three bonds should be formed and each HNH angle should be 90°, i.e., ammonia should be a pyramidal and not a planar molecule. The NH_3 molecule is indeed pyramidal and the observed HNH angle is 107.3°. The actual bond angle is again larger than that predicted by the theory.

It might be argued that since the N atom possesses a half-filled $2p$ shell (its electronic configuration is $1s^22s^22p^3$), its density distribution is spherical and hence the N atom should not exhibit any directional preferences in its bonding. This argument is incorrect for the following reason. The density distribution is obtained by squaring the wave function. The wave function which properly describes the system must be obtained first, then squared to obtain the density. The wave function which describes the ammonia molecule consists of products of hydrogen $1s$ orbital functions with the nitrogen $2p$ orbital functions. (A product of orbitals is the mathematical statement of the phrase "overlap of orbitals" in valence bond theory.) The density distribution obtained by squaring the product of two orbitals is *not* the same as that obtained from the sum of the squares of the individual orbitals. Thus in the valence bond theory of molecular electronic structure the directional properties of the valence orbitals play an important role. By assuming that the most stable bond results when the two nuclei joined by the bond lie along the axis defined by the orbitals and considering the bonds to a first approximation to be independent of one another, we can predict the geometries of molecules.

Hybridization

The BeH_2 molecule is linear and the two *Be—H* bonds are equivalent. The valence bond description of BeH_2 accounted for the two-fold valency of *Be* (which has the ground state configuration $1s^2 2s^2$) by assuming the bonding to occur with a promoted configuration of *Be*:

Be* $1s^2 2s(\uparrow) 2p(\uparrow)$

At first sight this suggests that the two *Be—H* bonds should be dissimilar and not necessarily 180° apart because one bond results from the overlap with a 2*s* orbital and the other with a 2*p* orbital on Be. We can, however, account for the equivalence of the two *Be—H* bonds and for the linearity of the molecule within the framework of the theory. There is no *a priori* reason for assuming that the one bond will result from the overlap with a 2*s* orbital and the other from the overlap with a 2*p* orbital. In the most general treatment of the problem, each bond to a hydrogen could involve both the 2*s* and the 2*p* orbitals. That is, we can "mix" or *hybridize* the valence orbitals on the Be atom. In fact, by taking each valence orbital on *Be* to be an equal part of 2*s* and 2*p*, we can obtain two equivalent hybrid orbitals which are directed 180° apart. The two hybrid orbitals will form two equivalent bonds with the *H* 1*s* orbitals whose total bond strength will be larger than that obtained by forming one bond with a 2*p* and the other with a 2*s* orbital on Be.

The construction of the hybrid orbitals is accomplished by taking the sum and the difference of the 2*s* orbital and one of the 2*p* orbitals, say the $2p_x$ orbital, both orbitals being centred on the Be nucleus.

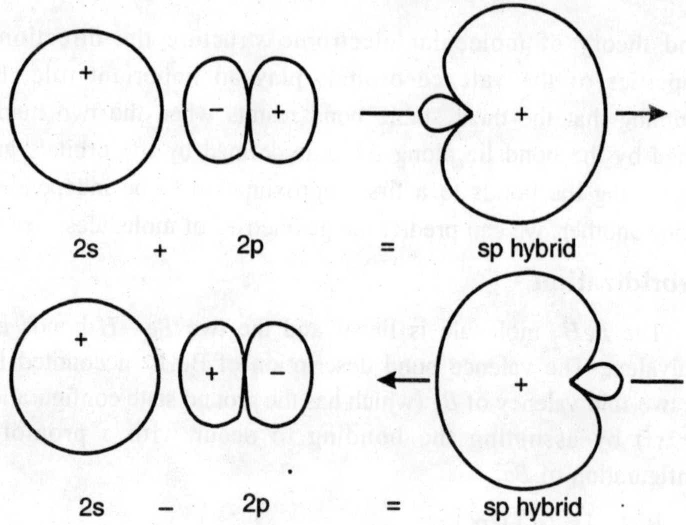

Fig. The construction of *sp* hybrid orbitals from a 2*s* and 2*p* atomic orbital on *Be*.

Since the 2*p* orbital has a node at the nucleus the 2*p* orbital wave function has opposite signs on each side of the nodal plane indicated in the figure. Both orbitals are positive on one side and the orbital functions add at each point in space. On the other side of the nodal plane, the orbitals are of opposite sign and their sum yields the difference between the two functions at every point in space. The addition of a 2*s* and $2p_x$ orbital concentrates the wave function and hence the charge density on the positive side of the *x*-axis. Obviously the combination ($2s$–$2p_x$) will be similar in appearance but concentrated on the negative side of the *x*-axis. These combinations of the 2*s* and 2*p* orbitals yield *two* hybrid orbitals which are equivalent and oppositely directed. Since each of the hybrid orbitals is constructed from equal amounts of the 2*s* and 2*p* orbitals they are termed "*sp* hybrid" orbitals.

The linear nature of BeH_2 can be explained if it is assumed (as is true) that the best overlap with both H 1*s* orbitals will result when the valence orbitals on the Be are *sp* hybrids.

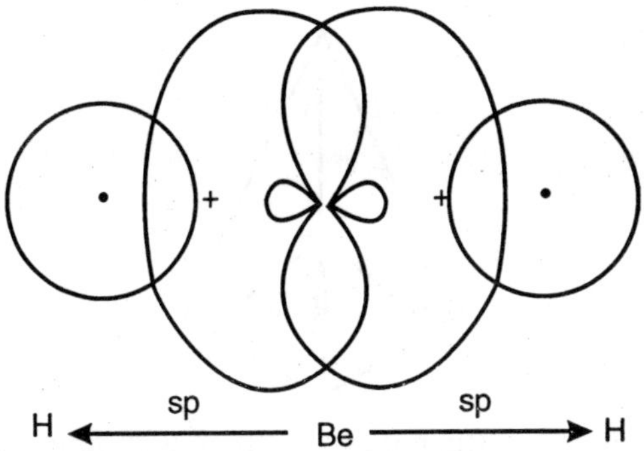

Fig A pictorial representation of the overlap of two *sp* hybrid orbitals on *Be* with *H* 1*s* orbitals to form BeH_2.

The three *B—H* bonds in BH_3 are equivalent and the molecule is planar and symmetrical:

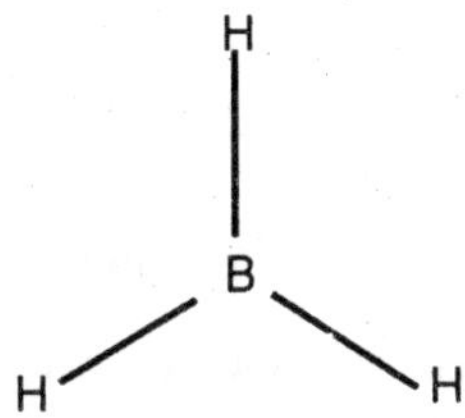

The promoted configuration of boron with three unpaired electrons is

B* $1s^2 2s(\uparrow) 2p_x(\uparrow) 2p_y(\uparrow)$

In this case we must construct three equivalent hybrid orbitals from the three atomic orbitals 2*s*, $2p_x$, and $2p_y$, on boron. The $2p_x$ and $2p_y$ orbitals define a plane in space and the three hybrid orbitals constructed from them will be projected in this same plane. Since the hybrid orbitals are to be equivalent, each must contain one part 2*s* and two parts 2*p*. They will be called "sp^2" hybrid orbitals. The three orbital combinations which have the above properties are indeed directed at 120° to one another. The planar, symmetrical geometry of BH_3 can be accounted for in terms of sp^2 hybridization of the orbitals on boron.

The four *C—H* bonds in CH_4 are equivalent and the molecule possesses a tetrahedral geometry:

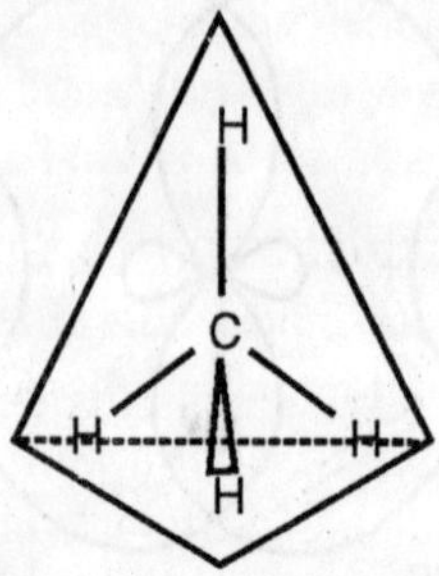

The promoted configuration of carbon with four unpaired spins is C* $1s^2 2s(\uparrow) 2p_x(\uparrow) 2p_y(\uparrow) 2p_z(\uparrow)$

Four equivalent hybrid orbitals can be constructed from the 2*s* and the three 2*p* orbitals on carbon. Each orbital will contain one part 2*s* and three parts 2*p*, and the hybrids are termed sp^3 hybrids. Only one such set of orbitals is possible and the angle between the orbitals is 109°28′, the tetrahedral angle. The tetrahedral geometry of CH_4 is described as resulting from the sp^3 hybridization of the valence orbitals on the carbon atom.

The three hybridization schemes which have been presented are sufficient to account for the geometries of all the compounds formed from elements of the first two rows of the periodic table (those with $n= 1$ or $n = 2$ valence orbitals). Consider, for example, the unsaturated hydrocarbons. The ethylene molecule, C_2H_4, possesses the planar geometry indicated here,

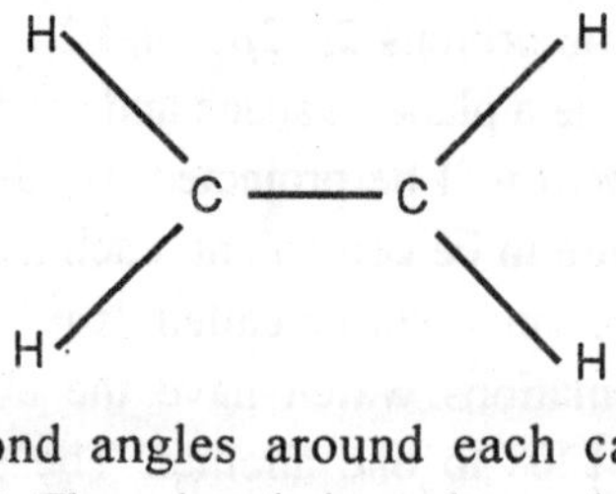

where the bond angles around each carbon nucleus are approximately 120°. Three bonds in a plane with 120° bond angles suggests sp^2 hybridization for the carbon atoms. Two of the sp^2 hybrids from each carbon may overlap with *H* 1*s* orbitals forming the four *C—H* bonds. The remaining sp^2 hybrids on each carbon may overlap with one another to form a bond between the carbons:

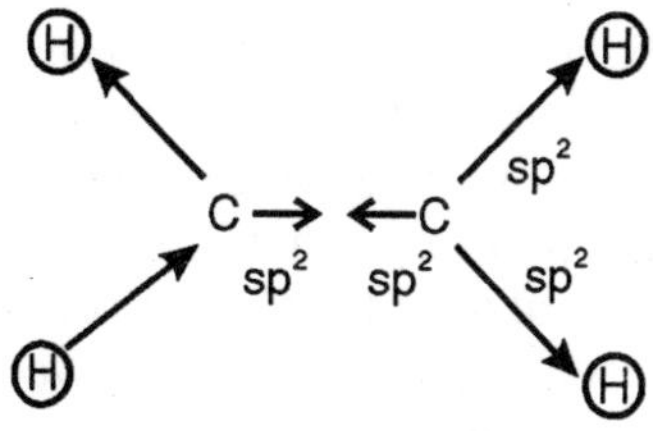

The sp^2 hybrids are denoted by arrows in the above diagram to indicate their directional dependence. If these bonds are formed in the *x-y* plane, using the $2p_x$ and $2p_y$ orbitals of the carbon atoms, a singly-occupied $2p_z$ orbital will remain on each carbon. They will be directed in a plane perpendicular to the plane of the molecule:

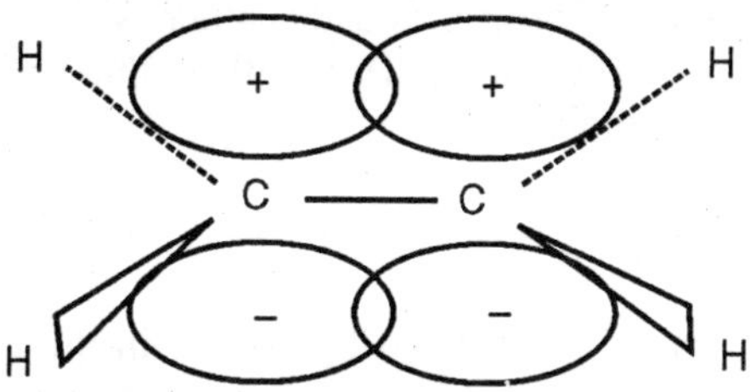

The overlap of the two $2p_z$ orbitals above and below the plane of the molecule will result in a second electron pair bond between the carbon atoms. The bonds formed in the plane of the molecule are called σ (sigma) bonds, while those perpendicular to the plane are called π bonds. Since the overlap of the orbitals to form a π bond is not as great as the overlap obtained from σ bonds (which are directed along the bond axis), π bonds in general are weaker than σ bonds. A Lewis structure for the C_2H_4 molecule is expressed as

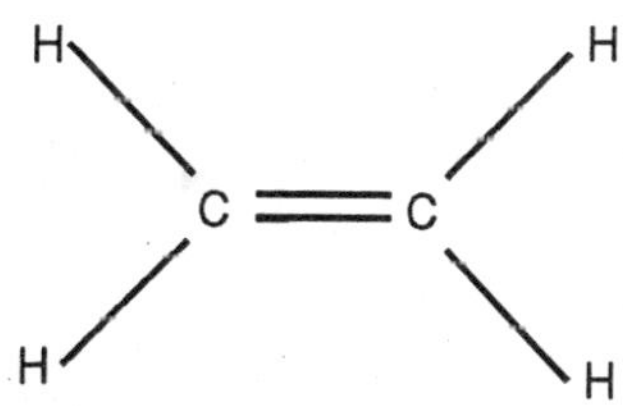

indicating that there is a double bond between the carbon atoms, i.e., the density from two pairs of electrons binds the carbon atoms.

The energy required to break the carbon-carbon double bond in ethylene is indeed greater than that required to break the carbon-carbon single bond in the ethane molecule, $H_3C—CH_3$. Furthermore, the chemical behaviour of ethylene is readily accounted for in terms of a model which places a large concentration of negative charge density in the region between the carbon atoms. The physical evidence thus verifies the valence bond description of the bonding between the carbons in ethylene.

Our final example concerns another important possible hybridization for the carbon atom. The acetylene molecule, C_2H_2, is a linear symmetric molecule: *H—C—C—H*. The linear structure suggests we try *sp* hybridization for each carbon, one hybrid overlapping with a hydrogen and the other with a similar hybrid from the second carbon atom. This will produce a linear σ bond framework for the molecule:

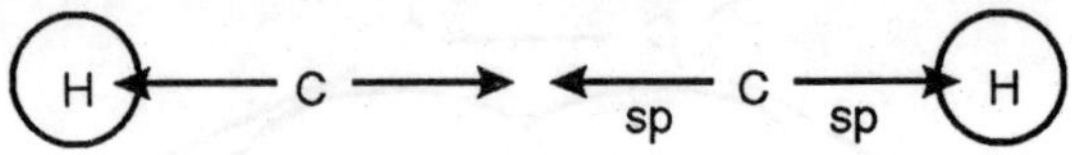

The *sp* hybrids are denoted by arrows in the above diagram. If the *sp* hybrids are assumed to be directed along the *x*-axis, then the remaining singly-occupied $2p_y$ and $2p_z$ orbitals on each carbon may form π bonds. The $2p_y$ orbitals on each carbon may overlap to form a π bond whose density is concentrated in the *x-y* plane, with a node in the *x-z* plane. Similarly the $2p_z$ orbitals may form a second π bond concentrated in the *x-z* plane, with a node in the *x-y* plane. Acetylene will possess a triple bond, one involving three pairs of electrons, between the carbon atoms. The Lewis structure is drawn as

where it is understood that one of the *C—C* bonds is a s bond while the other two are of the p-type. The chemistry and properties of acetylene are consistent with a model which places a large amount of charge density in the region of the *C—C* bond.

Hybridization schemes involving *d* orbitals are also possible. They are important for elements in the third and succeeding rows of the periodic table. Although the elements of the third row do not possess occupied 3*d* orbitals in their ground electronic configurations, the 3*d* orbitals of phosphorus, sulphur and chlorine are low enough in energy that promoted configurations involving the $3d$ orbitals may be reasonably postulated to account for the binding in compounds of these elements. One consequence of the "availability" of the $3d$ orbitals is that there are many exceptions to the octet rule in compounds of the third row elements. For example, in PCl_5 there are ten valence electrons involved in the bonding of the five chlorines to the phosphorus. A hybridization scheme based on the promotion of one $3s$ electron of phosphorus to a $3d$ orbital to yield five "dsp^3" hybrid orbitals correctly predicts the trigonal bypyramidal structure of PCl_5:

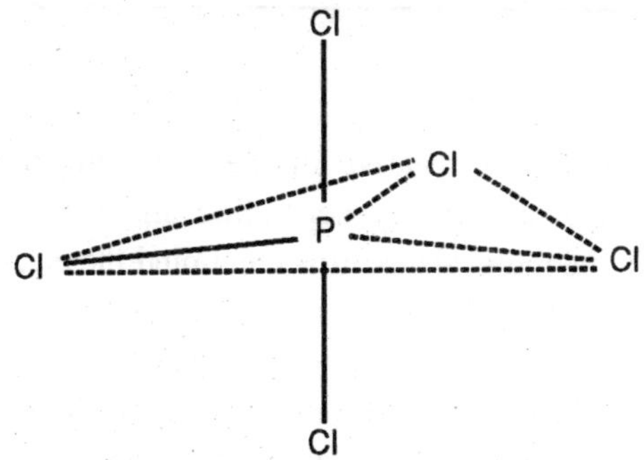

As a final example consider the molecule SF_6 in which all six S—F bonds are equivalent and the geometry is that of a regular octahedron (one F atom centred in each face of a regular cube):

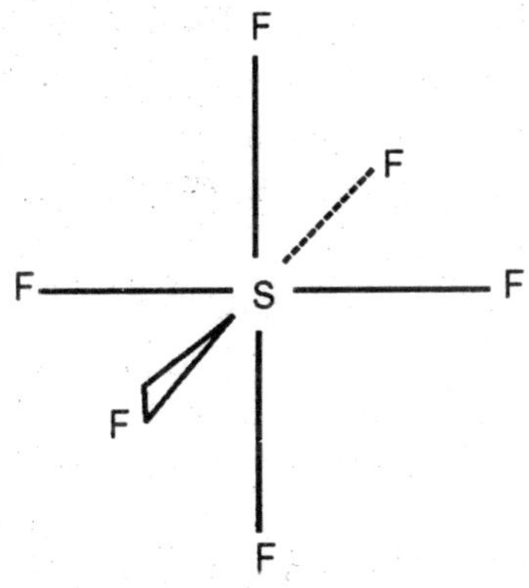

This geometry and number of bonds can be accounted for by assuming the promotion of one $3s$ and one $3p$ electron to two of the $3d$ orbitals on the sulphur atom. This hybridization yields six equivalent "d^2sp^3" hybrid bonds which are indeed directed as indicated in the structure for SF_6.

Chapter 5

Ionic and Covalent Binding

The distribution of negative charge in a molecule will exhibit varying degrees of asymmetry depending on the relative abilities of the nuclei in the molecule to attract and bind the electronic charge density. The symmetry or asymmetry of the charge distribution plays a fundamental role in determining the chemical properties of the molecule and consequently this property of the charge distribution is used as a basis for the classification of chemical bonds.

We can envisage two extremes for the distribution of the valence charge density. An example of one of the extremes is obtained when a bond is formed between two identical atoms. The charge density of the valence electrons will in this case necessarily be delocalized equally over corresponding regions of each nucleus since both nuclei will attract the electron density with equal force. Such an equal *sharing of the charge* density is an example of *covalent binding* and is exemplified by the molecular charge distribution of N_2.

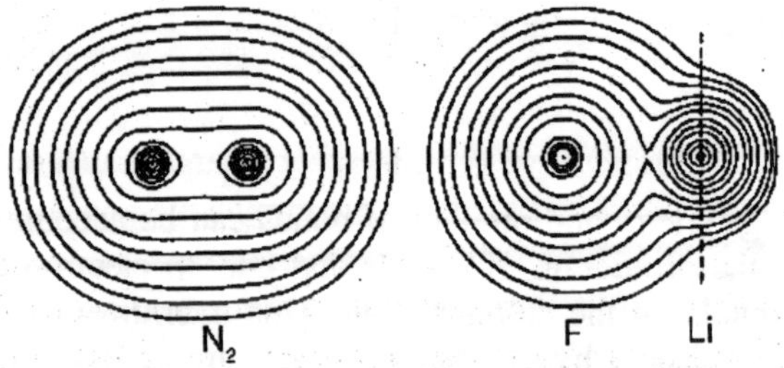

Fig. Contour maps of the molecular charge distributions of N_2 and LiF

The charge distribution of *LiF* provides an example of the other extreme, termed *ionic binding*, obtained when a bond is formed between two atoms with very different affinities for the electronic charge density. The very unsymmetrical distribution of charge in *LiF* is not simply a reflection of the fluorine atom possessing seven valence electrons to lithium's one. Instead the formation of the bond in *LiF* corresponds to the nearly complete transfer of the valence charge density of lithium to fluorine resulting in a molecule best described as Li^+F^-. We need only recall that initially a lithium atom is considerably larger than a fluorine atom to realise that a considerable transfer of charge has occurred in the formation of the *LiF* molecule.

In N_2 the valence charge density is delocalized over the whole molecule. The electronic charge is heavily concentrated in the internuclear region where it forms a bridge of high density between the two nuclei. Only the density of the 1*s* inner shell or "core" orbitals is strongly localized in the regions of the nuclei. In contrast to this, practically all of the charge density in the lithium fluoride molecule is localized in nearly spherical contours on the two nuclei in the manner characteristic of two separate closed-shell distributions. Only contours of very small value encompass both nuclei and the bridge of charge density joining the two spherical distributions is very low in value, being approximately one tenth of the value observed for N_2.

We may determine the total amount of electronic charge in an arbitrary region of space by summing the density in each small volume element within the region of interest (i.e., integrating the charge distribution over some particular volume of space):

$$\text{charge}\times\text{volume}=\left(e^{-}/a_0^3\right)\times a_0^3=e^{-}$$

A useful measure of the extent of charge transfer occurring on bond formation is obtained by determining the nonbonded charge on each nucleus. The nonbonded charge for a nucleus in a molecule is defined as occupying the volume of space on the nonbonded side of a plane perpendicular to the bond axis and through the nucleus in question. This is indicated by a dashed line for the *Li* nucleus in LiF.

The nonbonded charge density of the lithium nucleus in LiF is 1.07 e^- compared to 1.5 e^- in the *Li* atom (i.e., one half of the total number of electronic charges in a *Li* atom). The nonbonded charge density of the F nucleus, on the other hand, is increased above its atomic value, being 5.0 e^- as compared to 4.5 e^- in the fluorine atom. Since the distributions centred on the nuclei in LiF are nearly spherical, the total charge contained in each distribution will be approximately twice the value of the corresponding nonbonded charge. The total distribution of charge in *LiF* is, therefore, consistent with the ionic model Li^+F^- corresponding to the transfer of the single 2*s* valence electron of lithium to the fluorine. The radii of the nonbonded charge distributions (the distance measured along the bond axis from a nucleus to the outermost contour of its nonbonded density) are also consistent with the ionic model. The radius of the nonbonded charge density on lithium is $1.7a_o$, a value almost identical to the radius of a Li^+ ion (1.8 a_0) but much less than the radius of a *Li* atom (3.3 a_o). The value of 3.0 a_o for the radius of the nonbonded charge density on fluorine is consistent with that of a fluoride ion distribution as it represents a slight increase over the atomic value for fluorine of 2.8 a_o.

By way of comparison, the nonbonded charge on the nitrogen nuclei in N_2 is increased above the atomic value of 3.5 e^- to 3.68 e^-. This transfer of charge density to the nonbonded regions on bond formation is somewhat surprising when it is recalled that charge density must be accumulated in the binding region, the region between the nuclei, to achieve a chemical bond. We require a more detailed picture of the charge reorganization accompanying the formation of a bond to understand fully the distribution of the charge density in a molecule. In addition, many chemical bonds possess charge distributions which lie between the extreme of the perfect sharing of the valence charge density found in N_2 and its complete localization on one nucleus in *LiF*.

We consider next a method of classification of bonding in molecules, a classification which provides at the same time an understanding of the mechanism of the two binding situations in terms of the forces exerted on the nuclei.

Classification of Chemical Bonds

To make a quantitative assessment of the type of binding present in a particular molecule it is necessary to have a measue of the extent of charge transfer present in the molecule relative to the charge distributions of the separated atoms. This information is contained in the density difference or bond density distribution, the distribution obtained by subtracting the atomic densities from the molecular charge distribution. Such a distribution provides a detailed measure of the net reorganization of the charge densities of the separated atoms accompanying the formation of the molecule.

The density distribution resulting from the overlap of the undistorted atomic densities (the distribution which is subtracted from the molecular distribution) does not place sufficient charge density in the binding region to balance the nuclear forces of repulsion. The regions of charge increase in a bond density map are, therefore, the regions to which charge is transferred relative to the separated atoms to obtain a state of electrostatic equilibrium and hence a chemical bond. Thus we may use the location of this charge increase relative to the positions of the nuclei to characterize the bond and to obtain an explanation for its electrostatic stability.

In covalent binding we shall find that the forces binding the nuclei are exerted by an increase in the charge density which is shared mutually between them. In ionic binding both nuclei are bound by a charge increase which is localized in the region of a single nucleus.

Covalent Binding

The bond density map of the nitrogen molecule is illustrative of the characteristics of covalent binding.

The principal feature of this map is a large accumulation of charge density in the binding region, corresponding in this case to a total increase of one quarter of an electronic charge. As noted in the study of the total charge distribution, charge density is also transferred to the antibinding regions of the nuclei but the amount transferred to either region, 0.13 e^-, is less than is accumulated in the binding region. The charge density of the original atoms is decreased in regions perpendicular to the bond at the positions of the nuclei. In three dimensions, the regions of charge deficit correspond to two continuous rings or roughly doughnut-shaped regions encircling the bond axis.

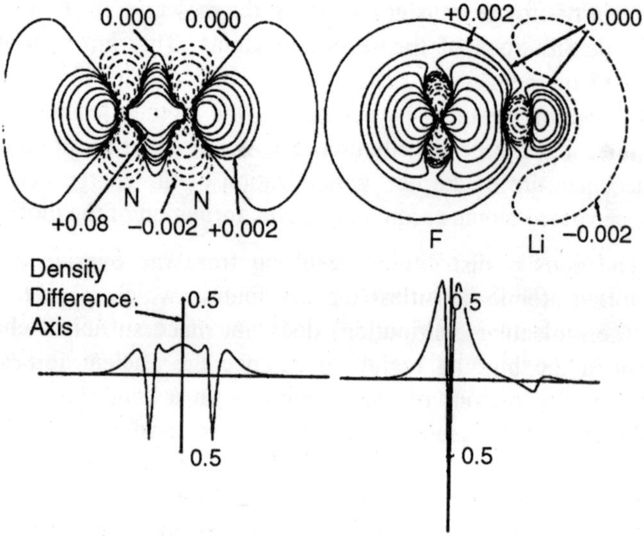

Fig. Bond density (or density difference) maps and their profiles along the internuclear axis for N_2 and *LiF*. The solid and dashed lines represent an increase and a decrease respectively in the molecular charge density relative to the overlapped atomic distributions. These maps contrast the two possible extremes of the manner in which the original atomic charge densities may be redistributed to obtain a chemical bond.

The increase in charge density in the antibinding regions and the removal of charge density from the immediate regions of the nuclei result in an increase in the forces of repulsion exerted on the nuclei, forces resulting from the close approach of the two atoms and from the partial overlap of their density distributions. The repulsive forces are obviously balanced by the forces exerted on the nuclei by the *shared* increase in charge density located in the binding region.

A bond is classified as covalent when the bond density distribution indicates that the charge increase responsible for the binding of the nuclei is shared by both nuclei. It is not necessary for covalent binding that the density increase in the binding region be shared equally as in the completely symmetrical case of N_2. We shall encounter heteronuclear molecules (molecules with different nuclei) in which the net force binding the nuclei is exerted by a density increase which, while shared, is not shared equally between the two nuclei.

The pattern of charge rearrangement in the bond density map for N_2 is, aside from the accumulation of charge density in the binding region, quite distinct from that found for H_2, another but simpler example of covalent binding. The pattern observed for nitrogen, a charge increase concentrated along the bond axis in both the binding and antibinding regions and a removal of charge density from a region perpendicular to the axis, is characteristic of atoms which in the orbital model of bonding employ p atomic orbitals in forming the bond. Since a p orbital concentrates charge density on opposite sides of a nucleus, the large buildup of charge density in the antibinding regions is to be expected.

In the orbital theory of the hydrogen molecule, the bond is the result of the overlap of s orbitals. The bond density map in this case is characterized by a simple transfer of charge from the antibinding to the binding region since s orbitals do not possess the strong directional or nodal properties of p orbitals. Further examples of both types of charge rearrangements or polarizations will be illustrated below.

Ionic Binding

We shall preface our discussion of the bond density map for ionic binding with a calculation of the change in energy associated with the formation of the bond in *LiF*. While the calculation will be relatively crude and based on a very simple model, it will illustrate that the complete transfer of valence charge density from one atom to another in forming a molecule is in certain cases energetically possible.

Lithium possesses the electronic configuration $1s^2 2s^1$ and is from group *IA* of the periodic table. It possesses a very low ionization potential and an electron affinity which is zero for all practical purposes. Fluorine is from group VIIA and has a configuration $1s^2 2s^2 2p^5$. It possesses a high ionization potential and a high electron affinity. The following calculation will illustrate that the $2s$ electron of Li could conceivably be transferred completely to the $2p$ shell of orbitals on F in which there is a single vacancy. This would result in the formation of a molecule best described as Li^+F^-, and in the electron configurations $1s^2$ for Li^+ and $1s^22s^22p^6$ for F^-.

We can calculate the energy change for the reaction

$$Li + F \rightarrow Li^+F^-$$

in stages. The energy which must be supplied to ionize the $1s$ electron on the Li atom is:

$$Li \rightarrow Li^+ + e^-, \ E_1 = I_1 = 5.4ev$$

The energy released when an electron combines with an F atom is given by the electron affinity of F:

$$F + e^- \rightarrow F^-$$

$$E_2 = -3.7ev$$

The two ions are oppositely charged and will attract one another. The energy released when the two ions approach one another from infinity to form the LiF molecule is easily estimated. To a first approximation it is simply $-e^2/R$ where R is the final equilibrium distance between the two ions in the molecule:

3. $Li^+ + F^- \rightarrow Li^+F^- \quad E_3 = -4ev\text{(approx)}$

(large distance apart) . (at R)

The sum of these three reactions gives $Li^+ + F^- \rightarrow Li^+F^-$

and the overall change in energy is the sum of the three energy changes, or approximately –2 ev. The species Li^+F^- possesses a lower energy than the separated Li and F atoms and should therefore be a stable molecule.

The transfer of charge density from lithium to fluorine is very evident in the bond density map for LiF. The charge density of the $2s$ electron on the lithium atom is a very diffuse distribution and consequently the negative contours in the bond density map denoting its removal are of large spatial extent but small in magnitude. The principal charge increase is nearly symmetrically arranged about the fluorine nucleus and is completely encompassed by a single nodal surface. The total charge increase on fluorine amounts to approximately one electronic charge. The charge increase in the antibinding region of the lithium nucleus corresponds to only 0.01 electronic charges. (The great disparity in the magnitudes of the charge increases on lithium and fluorine are most strikingly portrayed in the profile of the bond density map, also shown in Fig It is equally important to realise that the charge increase on lithium occurs within the region of the 1*s* inner shell or core density and not in the region of the valence density. Thus the slight charge increase on lithium is primarily a result of a polarization of its core density and not of an accumulation of valence density.

The pattern of charge increase and charge removal in the region of the fluorine, while similar to that for a nitrogen nucleus in N_2, is much more symmetrical, and the charge density corresponds very closely to the distribution obtained from a single $2p\sigma$ electron. Thus the simple orbital model of the bond in LiF which describes the bond as a transfer of the 2*s* electron on lithium to the single $2p\sigma$ vacancy on fluorine is a remarkably good one.

While the bond density map for *LiF* substantiates the concept of charge transfer and the formation of Li^+ and F^- ions it also indicates that the charge distributions of both ions are polarized. The charge increase in the binding region of fluorine exceeds slightly that in its antibinding region (the F^- ion is polarized towards the Li^+ ion) and the charge distribution of the Li^+ ion is polarized away from the fluorine. A consideration of the forces exerted on the nuclei in this case will demonstrate that these polarizations are a necessary requirement for the attainment of electrostatic equilibrium in the face of a complete charge transfer from lithium to fluorine.

Consider first the forces acting on the nuclei in the simple model of the ionic bond, the model which ignores the polarizations of the ions and pictures the molecule as two closed-shell spherical ions in mutual contact. If the charge density of the Li^+ ion is spherical it will exert no net force on the lithium nucleus. The F^- ion possesses ten electrons and, since the charge density on the F^- ion is also considered to be spherical, the attractive force this density exerts on the *Li* nucleus is the same as that obtained for all ten electrons concentrated at the fluorine nucleus. Nine of these electrons will screen the nine positive nuclear charges on fluorine from the lithium nucleus. The net force on the lithium nucleus is, therefore, one of attraction because of the one excess negative charge on *F*.

For the molecule to be stable, the final force on the lithium nucleus must be zero. This can be achieved by a distortion of the spherical charge distribution of the Li^+ ion. If a small amount of the 1*s* charge density on lithium is removed from the region adjacent to fluorine and placed on the side of the lithium nucleus away from the fluorine, i.e., the charge distribution is polarized away from the fluorine, it will exert a force on the lithium nucleus in a direction away from the fluorine. Thus the force on the lithium nucleus in an ionic bond can be zero only if the charge density of the Li^+ ion is polarized away from the negative end of the molecule.

A similar consideration of the forces exerted on the fluorine nucleus demonstrates that the F^- ion density must also be polarized. The fluorine nucleus experiences a net force of repulsion because of the presence of the lithium ion. The two negative charges centred on lithium screen only two of its three nuclear charges. Therefore, the charge density of the F^- ion must be polarized *towards* the lithium in order to exert an attractive force on the fluorine nucleus which will balance the repulsive force arising from the presence of the Li^+ ion. Thus both nuclei in the LiF molecule are bound by the increase in charge density localized in the region of the fluorine.

The charge distribution of a molecule with an ionic bond will necessarily be characterized not only by the transfer of electronic charge from one atom to another, but also by a polarization of each of the resulting ions in a direction counter to the transfer of charge, as indicated in the bond density map for *LiF*.

The bond density maps for N_2 and *LiF* are shown side by side to provide a contrast of the changes in the atomic charge densities responsible for the two extremes of chemical binding. *In a covalent bond the increase in charge density which binds both nuclei is shared between them. In an ionic bond both nuclei are bound by the forces exerted by the charge density localized on a single nucleus.* It must be stressed that there is no fundamental difference between the forces responsible for a covalent or an ionic bond. They are electrostatic in each case.

MOLECULAR CHARGE DISTRIBUTIONS OF HOMONUCLEAR DIATOMIC MOLECULES [H]

Contour maps of the charge distributions for the stable homonuclear diatomic molecules formed from the second-row atoms provide further examples of covalent binding. The maps illustrate the relative tightness of binding of the density distributions, the density in Li_2 for example being much more diffuse than that in N_2. Two important physical dimensions for a molecule are the bond length and the molecular size. The bond length of a molecule may be directly determined (by *X*-ray diffraction techniques or by spectroscopic methods) but the size of a molecule cannot be as precisely defined or measured. However, molecular diameters may be inferred from measurements of the viscosity of gas phase molecules and from *X*-ray crystallographic studies on the structures of molecular crystals such as solid N_2 and O_2.

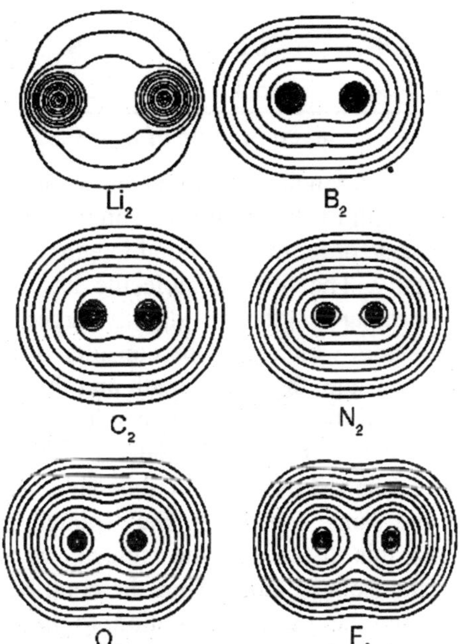

Fig. Countour maps of the molecular charge distribution for the stable homonuclear diatomic molecules Li_2 to F_2.

In general over 95 per cent of the molecular charge lies within the 0.002 contour and it has been found that the dimensions of this contour agree well with the experimental estimates of molecular sizes. The length and width of each molecule, defined respectively as the distance between the intercepts of the 0.002 contour on the molecular axis and on a line perpendicular to the axis and passing through its mid-point, are given in Table along with the experimental bond lengths R_e

Table. Properties of the Total Charge Distributions*

A_2	R_e	Length	Width	Nonbonded Radius	
				Molecule	Atom
Li_2	5.051	8.7	7.8	1.8	3.3
B_2	3.005	9.8	7.2	3.4	3.4
C_2	2.3481	8.5	7.0	3.1	3.2
N_2	2.068	8.2	6.4	3.1	3.0
O_2	2.282	7.9	6.0	2.8	2.9
F_2	2.68	7.9	5.4	2.6	2.8

*All distances are given in units of $a_o = 0.52917$ Å.

There is only a rough correlation between the bond length and the overall length of the molecule. Thus the lengths of N_2 and O_2 are in the reverse order of their bond lengths, as is also roughly true experimentally. The lithium molecule has the largest bond length but a molecular length only slightly larger than that of C_2. There are two factors which must be considered in understanding the length of a molecule, the bond length and the rate at which the density falls off from the nucleus on the side away from the bond. Table lists the distance from the nucleus to the 0.002 contour in the molecule, i.e., the radius of the nonbonded charge density, and the radius of the same contour in the isolated atom. With the exception of Li_2, this distance in the molecule is almost identical to the value in the isolated atom. Thus the contribution of the two end lengths, beyond the nuclear separation, to the overall length of a molecule is largely determined by how tightly the density is bound in the unperturbed atom. The binding of the atomic densities increases from *Li* across to *F*, so that *Li* and *Be* are large and diffuse and *N*, 0, and *F* progressively tighter and more compact. Therefore F_2 is smaller in size than N_2 or C_2 even though it possesses a greater bond length because the density in the F atom is more tightly bound than that in the *C* or *N* atoms. The *Li* molecule differs from the others in that its length is considerably less than expected considering the diffuse nature of its atomic density. In this case the molecular length is not approximately equal to the sum of R_e and twice the "atomic" radius. This is, however, easily understood since in the *Li* atom only one valence-shell electron is present and in the molecule the charge density of this electron is concentrated almost exclusively in the binding region. This is further illustrated by using instead of the 0.002 contour of *Li* the 0.002 contour of the $1s^2$ shell of Li^+, which is in fact equal to the value listed in Table for the Li_2 molecule.

An estimate of the size of a peripheral atom in a molecule can thus be obtained by taking the sum of $\frac{1}{2}R_e$ from a suitable source and the atomic radius as defined by the 0.002 contour of the atom.

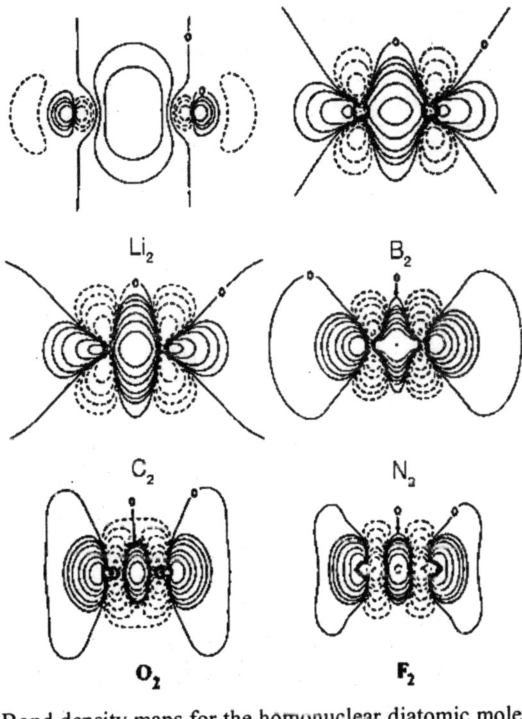

Fig. Bond density maps for the homonuclear diatomic molecules.

The bond density maps for the second-row homonuclear diatomic molecules indicate that the original atomic densities are distorted so as to place charge in the antibinding as well as in the binding regions.

Apart from Li_2 the pattern of charge increase and charge removal in these molecules is similar to that discussed previously for N_2, a pattern ascribed to the participation of $2p\sigma$ orbitals in the formation of the bond. Only Li_2 approximates the simple picture found for H_2 of removal of charge from the antibinding region and a buildup in the binding region. For the remaining molecules charge density is increasingly accumulated along the bond axis in both the binding and antibinding regions.

The total accumulation of electronic charge represented by the regions of positive contours in the binding and antibinding regions of the bond density maps are listed in Table.

Table. Charge Contained in the Regions of Increase in Bond Density Maps

A_2	Binding Region	Antibinding Region
Li_2	0.41	0.01
B_2	0.30	0.05
C_2	0.50	0.06
N_2	0.25	0.13
O_2	0.10	0.14
F_2	0.08	0.10

These figures show that in O_2 and F_2 a greater amount of charge is transferred to the antibinding region of a single nucleus than to the binding region. It is evident, however, from the shapes of the contours that the charge increase in the binding region is concentrated along the bond axis where it exerts a maximum force of attraction on the nuclei while the buildup in the antibinding region is more diffuse.

The net forces on the nuclei are zero for each molecule. Therefore, the force exerted by the charge density in the binding region balances not only the force of nuclear repulsion but the force exerted by the charge buildup in the antibinding region as well. The nuclei are in each case bound by the charge increase which is shared equally by both nuclei.

An important physical property of a molecule is its bond energy, the amount of energy required to break the bond or bonds in a molecule and change it back into its constituent atoms. The bond energies of the second-row homonuclear diatomic molecules increase from either Li_2 or F_2 to a maximum value for the central member of the series, N_2.

Table. Bond Energies for Homonuclear Diatomic Molecules

Molecule	Bond Energy (ev)	Number of electron pair bonds
Li_2	1.12	1
B_2	3.0	1
C_2	6.36	2
N_2	9.90	3
O_2	5.21	2
F_2	1.65	1

We may rationalize the variation in the bond energies and the differences in the bond density maps in terms of the orbital theory of bonding. The valency of an atom to its number of unpaired electrons. Thus the number of electron pair bonds formed between atoms in this series of molecules is predicted to be one for Li_2, B_2 and F_2, two for C_2 and O_2, and three for N_2. Table reveals a parallelism between the bond energy and the number of electron pair bonds present in each molecule.

The detailed variation in bond energy through the series can be accounted for in terms of the type of bond (whether it is formed for *s* or *p* orbitals) present in each molecule, a feature which is clearly reflected in the bond density maps, and even more strikingly portrayed in their profiles.

The bond in Li_2 is formed primarily from the overlap of $2s$ atomic orbitals on each lithium atom. The $2s$ atomic density of lithium is a diffuse spherical distribution. These same characteristics are evident in the total charge distribution for Li_2 and particularly in its bond density map. The charge increase in the binding region, while large in amount, is very diffuse and the bond density profile shows that relative to the other molecules, the charge increase is not concentrated along the bond axis. These are the very features expected for a bond resulting from the overlap of distorted, *nondirectional s* orbitals.

B_2 and F_2 also have but a single pair bond. However, the bonds in these two molecules are formed primarily from the overlap of $2p\sigma$ orbitals. Since a $p\sigma$ orbital is directed along the bond axis, it is more effective than an s orbital at concentrating charge density along this same axis. This is particularly evident when we compare the profiles of the bond densities for F_2 and B_2 with the profile for Li_2. Similarly, the presence of two electron pair bonds and the still larger bond energies found for C_2 and O_2 are reflected in the larger increases in the charge densities along the internuclear axis in the binding region. Notice that while B_2 concentrates three times as much charge as O_2 in the binding region, it is not concentrated along the bond axis to as great an extent as in O_2, and consequently its bond energy is the smaller of the two.

The nitrogen molecule possesses three electron pair bonds and the largest bond energy of the molecules in this series. The charge increase in the binding region is concentrated along the bond axis to a far greater extent in this molecule than in any of the other molecules in the series. This concentration of the charge density gives N_2 a stronger bond than C_2 even though the total charge increase in its binding region is only one half as great as that for C_2.

The comparison of the bond energies in this series of molecules clearly illustrates that the strength of a bond is not simply related to the number of electronic charges in the binding region. As important as the *amount* of charge is the exact disposition of the charge density in the molecule, whether it is diffuse or concentrated.

DIPOLE MOMENTS AND POLAR BONDS [H]

Any chemical bond results from the accumulation of charge density in the binding region to an extent sufficient to balance the forces of repulsion. Ionic and covalent binding represent the two possible extremes of reaching this state of electrostatic equilibrium and there is a complete spectrum of bond densities lying between these two extremes. Since covalent and ionic charge distributions exhibit radically different chemical and physical properties, it is important, if we are to understand and predict the bulk properties of matter, to know which of the two extremes of binding a given molecule most closely approximates.

We can obtain an experimental measure of the extent to which the charge density is unequally shared by the nuclei in a molecule. The physical property which determines the asymmetry of a charge distribution is called the dipole moment. To illustrate the definition of the dipole moment we shall determine this property for the *LiF* molecule assuming that one electron is transferred from *Li* to *F* and that the charge distributions of the resulting ions are spherical.

The dipole moment is defined as the product of the total amount of positive or negative charge and the distance between their centroids The centroids of the positive and negative charges in a molecule are determined in a manner similar to that used to determine the centre of mass of a system.

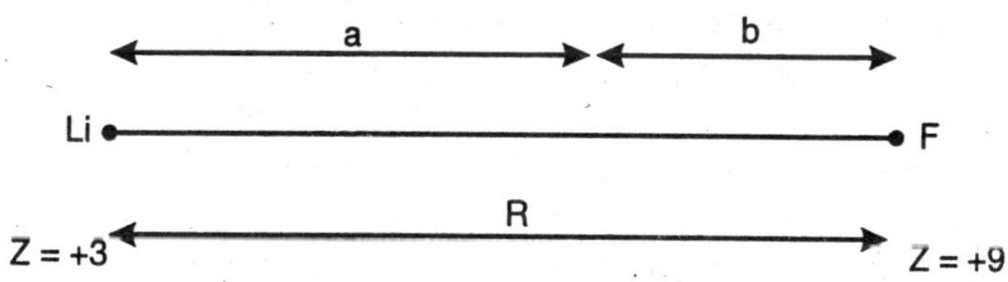

Fig. Diagram for The Calculation of The Centroids of Positive and Negative Charge in *LiF*.

With reference to Fig the "centre of gravity" of the positive charge in *LiF* is easily found from the following equations:

$$3a = 9b$$

$$a+b = R$$

Eliminating b from these equations and solving for a we find that

$$a = (3/4)$$

Thus all the positive charge in the LiF molecule can be considered to be at a point one quarter of the bond length away from the fluorine nucleus. Similarly the centroid of negative charge, remembering that one electron has been transferred from Li to F, is found to lie at a point one sixth of the bond length away from the F nucleus. The centroids of positive and negative charge do not coincide, the negative centroid being closer to the F nucleus than the positive centroid. While the molecule is electrically neutral, there is a separation of charge within the molecule. Let us denote the distance between the centroids of charge by l:

$$\begin{aligned} l &= (5/6-3/4)R \\ &= (1/12)R \end{aligned}$$

and since there are twelve electrons in LiF, the dipole moment denoted by μ is

$$\begin{aligned} \mu &= (12e)(1/12R) \\ &= eR \end{aligned}$$

Thus, not surprisingly, the dipole moment in this case is numerically equal to one excess positive charge at the Li nucleus and one excess negative charge at the F nucleus, or one pair of opposite charges separated by the bond length.

We can easily calculate the value of the dipole moment. The value of R for LiF is 1.53×10^{-8} cm and the charge on the electron is 4.80×10^{-10} esu. Thus

$$\begin{aligned} \mu &= eR \\ &= 7.34\times10^{-18} \\ \text{or } \mu &= 7.34 \text{debyes} \end{aligned}$$

where 1 debye = 1×10^{-18} esu cm. (The fundamental unit for dipole moments is called a debye in honour of *P*. Debye who was responsible for formulating the theory and method of measurement of this important physical quantity.) The experimental value of μ for *LiF* is slightly smaller than the calculated value, being 6.28 debyes. The reason for the discrepancy is easily traced to the assumption made in the calculation that the charge distributions of the Li^+ and F^- ions are spherical. We have previously indicated that the charge distributions of both the F^- and Li^+ ions are polarized in a direction counter to the direction of transfer of the electron in order to balance the forces on the nuclei. The centroid of the ten negative charges on F is not at the F nucleus, but shifted slightly towards the *Li*, and the centroid of the charge density on Li^+ is correspondingly shifted slightly off the *Li* nucleus away from the *F*. Thus the centroid of negative charge for the whole molecule is not as close to the *F* nucleus as our simple calculation indicated and the dipole moment is correspondingly less.

Obviously from this discussion the dipole moment of a molecule with a covalent bond will be zero since the symmetry of the charge distribution will dictate that the positive and negative charge centroids coincide. Thus dipole moments can conceivably possess values which lie between the covalent limit of zero and the ionic extreme which approaches *neR* in value (*n* being the number of electrons transferred in the formation of the ionic bond).

The series of diatomic molecules formed by the union of a single hydrogen atom with each of the elements in the second row of the periodic table exemplifies both the extreme and intermediate types of binding, and hence of dipole moments. Table lists the dipole moments and the values of *eR* for the ionic extreme (assuming spherical ions) for the second-row diatomic hydride molecules.

Table.

Dipole Moments and Bond Lengths of Second-row Hydrides

AH	m*(debyes)	eR(debyes)	R(Å)
LiH	-6.002	-7.661	1.595
BeH	-0.282	-6.450	1.343

BH	1.733	5.936	1.236
CH	1.570	5.398	1.124
NH	1.627	4.985	1.038
OH	1.780	4.661	0.9705
FH	1.942	4.405	0.9171

*The negative or positive signs for μ imply that H is the negative or positive end of the dipole respectively.

All of these molecules exist as stable, independent species in the gas phase at low pressures and may be studied by spectroscopic methods or by molecular beam techniques. Only *LiH* and *HF*, however, are stable under normal conditions; *LiH* is a solid and *HF* a gas at room temperature. The remaining diatomic hydrides are very reactive since they are all capable of forming one or more additional bonds.

The variation of the dipole moment in this series of molecules provides a measure of the *relative abilities* of *H* and of each of the second-row elements to attract electrons. For example, the dipole moment for *LiH* illustrates that electron density is transferred from *Li* to *H* in the formation of this molecule. In *HF*, on the other hand, charge density is transferred from *H* to *F*. With the exception of *BH*, there is a steady increase in μ from -6.0 debyes for *LiH* to +1.9 debyes for *HF*. Only *LiH* approaches the ionic limit of Li^+H^-. *BeH* appears to possess a close to equal sharing of the valence electrons. The remaining molecules, while exhibiting some degree of charge removal from *H*, are all far removed from the ionic extreme. They represent cases of molecular binding which lie between the two extremes, ionic and covalent. They are referred to as polar molecules.

We can best illustrate the variation in the chemical binding in this series of molecules by considering the properties of the molecular charge and bond density distributions. In *LiH* almost all of the molecular charge density is centred on the two nuclei in nearly spherical distributions.

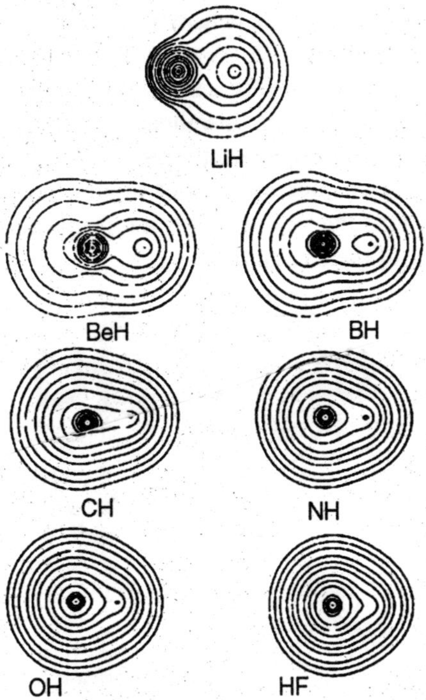

Fig. Contour maps of the molecular charge distrubution of the diatomic hydride molecules LiH to HF. The proton is the nucleus on the right-hand side in each case.

The nonbonded charge and radius for lithium, 1.09 e^- and 1.7 a_0 respectively, are characteristic of the $1s^2$ inner shell distribution of Li^+. Thus the molecular charge distribution for *LiH* indicates that the single valence electron of lithium is transferred to hydrogen and that the bond is ionic.

In *BeH*, the valence density has the appearance of being equally shared by the two nuclei. From *BH* through to *HF* a decreasing amount of density is centred on the proton to the extent that the charge distribution of *HF* could be approximately described as an F^- ion distribution polarized by an imbedded proton.

The increase of the effective nuclear charge across a row of the periodic table is reflected not only in the amount of charge transferred to or from the hydrogen, but also in the relative sizes of the molecules. In BeH the density is diffuse and the molecule is correspondingly large. In *HF* the density is more compact and the molecule is the smallest in the series. The decrease in the size of the molecule from *BH* to *HF* parallels the decrease in the size of the atoms *B* to *F*. The intermediate size of *LiH* is a consequence of the one and only valence electron of lithium being transferred to hydrogen, and thus the size of *LiH* is a reflection of the size of the Li^+ ion and not of the Li atom.

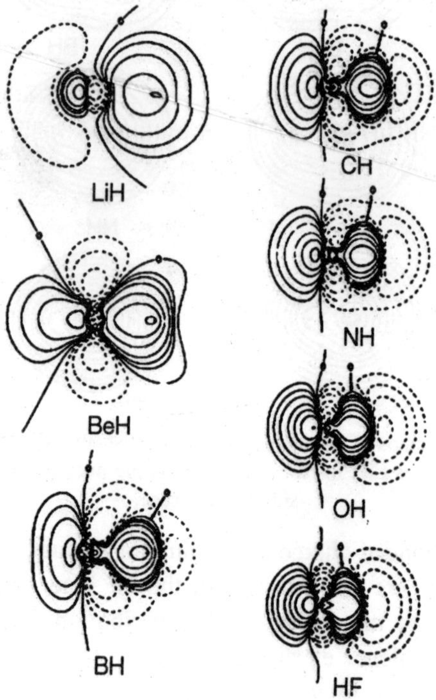

Fig. Bond Density Maps for The Diatomic Hydride Molecules LiH to HF.

In general terms, the bond density maps provide a striking confirmation of the transfer of charge predicted by the relative electron affinities or by the relative effective nuclear charges of hydrogen and the second-row elements *Li* → *F*. We may again employ the position of the charge increase in the bond density map to characterize the type of binding present in the molecule. The map for *LiH* exhibits the same characteristics as does the one for *LiF*, the contours in the region of the *Li* nucleus being remarkably similar in the two cases. The valence density is clearly localized about the proton just as it is about the fluorine nucleus in *LiF*. The $1s$ core density remaining on lithium is clearly polarized away from the proton, and the density increase localized on the proton is polarized towards the lithium as required in ionic binding.

The one principal difference between the *LiH* and *LiF* bond density maps concerns the shape of the contours representing the density increase on the proton and fluorine nucleus. In *LiF* the contours on fluorine are similar in shape to those obtained for a $2p\sigma$ orbital density. In *LiH* the contours on the proton are nearly spherical. In terms of a simple orbital model we imagine the $2s$ electron of *Li* to be transferred to the $1s$ orbital of hydrogen in *LiH* and to the $2p\sigma$ orbital of fluorine in *LiF*. The spherical and double-lobed appearance of the density increases found for hydrogen and fluorine respectively show these orbital models of the binding to be reasonable ones.

From *BeH* through the rest of the series, the bond density maps show an increase in the amount of charge removed from the proton and transferred to the region of the other nucleus. This is evident from the increase in the number and diameter of the dashed contours in the nonbonded region of the proton. The pattern of charge increase and charge removal in the regions of the *Be*, *B*, *C*, *N*, *O* and *F* nuclei is similar to that found for these nuclei in their homonuclear diatomic molecules, and is characteristic of the participation of a $p\sigma$ orbital in the formation of the bond. The polarization of the density in the region of the hydrogen is of the simple dipolar type characteristic of a dominant s orbital contribution. As previously discussed, the double-lobed appearance of the density increase in the region of fluorine in the bond density map for *LiF* can be viewed as characteristic of the ionic case when a $2p\sigma$ orbital vacancy is filled in forming the bond. This limiting pattern is most closely approached in the hydride series by *HF*, the molecule exhibiting the largest degree of charge transfer from hydrogen. *HF*, of all the hydrides, is most likely to approach the limiting ionic extreme of H^+F^-. However, the charge increase in the region of fluorine in *HF* is not as symmetrical as that found for *F* in the *LiF* molecule. The proton in *HF*, unlike the Li^+ ion in *LiF*, is imbedded in one lobe of the density increase on *F* and distorts it. Thus, unlike the ionic extreme of *LiF*, the charge increase on *F* in *HF* is shared by both nuclei in the molecule.

Another important difference between the charge distributions of *HF* and *LiF* concerns the polarizations of the charge density in the immediate vicinities of the nuclei. In *LiF* (or *LiH*) the localized charge distributions are both polarized in a direction *opposite* to the direction of charge transfer Li $\rightarrow$ F (or Li $\rightarrow$ H). These polarizations are a consequence of the extreme charge transfer from lithium to fluorine, a transfer resulting in a force of attraction on the lithium nucleus and one of repulsion on the fluorine nucleus. In *HF* the charge density in the regions of the proton and the fluorine nucleus is polarized in the *same* direction as the direction of charge transfer from $H \rightarrow F$. Thus the amount of charge transferred to the vicinity of the fluorine in *HF* is not, unlike the situation in *LiF*, sufficient to screen the nuclear charge of fluorine and hence exert a net attractive force on the proton. Instead, the fluorine nucleus and its associated charge density exert a net repulsive force on the proton, one which is balanced by the inwards polarization of the charge density in the region of the proton.

The polarization of the charge density on the proton adds to and is contiguous with the charge increase in the binding region. Thus in *HF* and in the molecules *BeH* to *OH* for which the charge transfer is less extreme, the nuclei are bound by a shared density increase and the binding is covalent. From *BeH* through the series of molecules the sharing of the charge increase in the binding region becomes increasingly unequal and favours the heavy nucleus over the proton. The latter molecules in the series, *NH*, *OH* and *HF*, provide examples of polar binding which are intermediate between the extremes of perfect covalent and ionic binding as exhibited by the homo nuclear diatomics and *LiF* respectively.

In general, chemical bonds between identical atoms or between atoms from the same family in the periodic table will exhibit equal or close to equal sharing of the bond density and be covalent in character. Compounds formed by the union of elements in columns I or II with elements in columns VIA or VIIA will be ionic, as exemplified by *LiF* or *BeO*. We find a continuous change from covalent to ionic binding as the atoms joined by a chemical bond come from columns in the periodic table which are progressively further removed from one another.

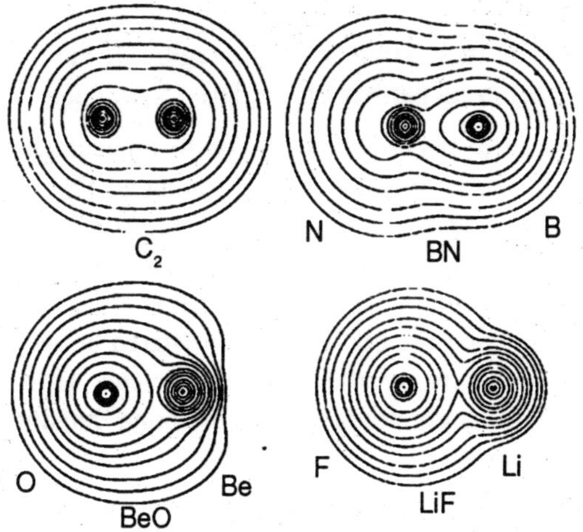

Fig. Molecular Charge Distributions for the 12-electron Isoelectronic series C_2, *BN*, *BeO* and *LiF*.

This series of molecules is formed (in an imaginary process) by the successive transfer of one nuclear charge from the nucleus on the left to the nucleus on the right, starting with the central symmetrical molecule C_2.

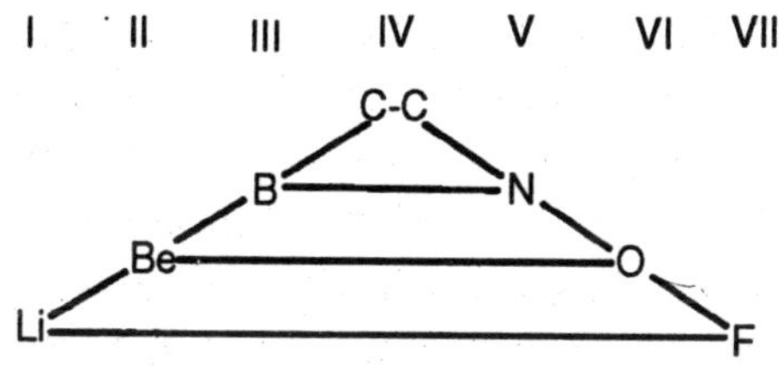

The molecules are said to form an isoelectronic series since they all contain the same number of electrons, twelve. The molecular charge distributions in this series illustrate how the charge distribution and binding for a constant number of electrons changes as the nuclear potential field in which the electrons move is made increasingly unsymmetrical.

In C_2 the nuclear charges are, of course, equal and the charge distribution is symmetrically shared by both nuclei in the manner characteristic of covalent binding. In the remaining molecules the valence charge density is increasingly localized in the region of the left-hand nucleus. This is particularly evident in the bond density maps and their profiles which show the increasing extent to which charge density is transferred from the region of the nucleus on the right (*B*, *Be*, *Li*) to its partner on the left (*N*, *0*, *F*).

The charge distribution of *BN* (with nuclear charges of five for boron and seven for nitrogen) is similar to that for C_2 in that charge is accumulated in the nonbonded regions of both nuclei as well as in the region between the nuclei. However, the buildup of charge behind the boron nucleus is smaller than that behind the nitrogen nucleus and the charge density shared between the nuclei is heavily shifted towards the nitrogen nucleus. Thus the binding in BN is predominantly covalent, but the bond density is polarized towards the nitrogen.

The charge transfer in *BeO* and *LiF* is much more extreme and the bond density maps show a considerable loss of charge from the nonbonded regions of both the *Be* and *Li* nuclei. Notice that except for contours of very low value the charge density in *BeO*, as in *LiF*, is localized in nearly spherical distributions on the nuclei, distributions which are characteristic of Be^{+2} and O^{-2} ions. A count of the number of electronic charges contamed within the spherical or nearly spherical contours centred on the nuclei in *BeO* and *LiF* indicates that the charge distributions correspond to the formulae $Be^{+1.5}O^{-1.5}$ and $Li^{+1}F^{-1}$. That is, the binding is ionic and corresponds to the transfer of approximately one charge from *Li* to *F* and of one and one half charges from *Be* to *O*. Thus while the binding in *LiF* is close to the simple orbital model of $Li^{+}(1s^2)F^{-}(1s^22s^22p^6)$ as noted before, the binding in *BeO* falls somewhat short of the description $Be^{+2}(1s^2)O^{-2}(1s^22s^22p^6)$. Notice that the density contours on oxygen in *BeO* are more distorted towards the *Be* than the contours on *F* are towards *Li* in *LiF*. This illustrates that the oxygen anion is more polarizable than is the fluoride anion.

The radius of the charge distribution on the nonbonded side of the *Be* nucleus as measured along the bond axis is identical to that found for an isolated Be^{+2} ion. However, the radius of the *Be* charge density perpendicular to the bond axis is much greater than that for a Be^{+2} ion. This shows, as does the actual electron count given above, that the two valence electrons of boron are not completely transferred to oxygen in the formation of the BeO molecule.

Hydrogen is an exception to the above set of generalizations regarding an element's position in the periodic table and the ionic-covalent nature of the bond it forms with other elements. It does not behave in a manner typical of family *IA*. The bond formed between hydrogen and another member of group *IA*, as exemplified by LiH, is ionic and not covalent. Here hydrogen accepts a single electron to fill the vacancy in its 1*s* shell and thus resembles the members of family VIIA, the halogens. The bond in HF, however, is more polar than would be expected for the union of two adjacent members of the same family, and hydrogen is therefore not a typical member of family VIIA. This intermediate behaviour for H is understandable in that the values of its ionization potential and electron affinity are considerably greater than those observed for the alkali metals (IA) but are considerably less than those found for the halogens (VIIA).

Electronegativity

It is important that we be able to predict the extent to which electronic charge will be transferred from one atom to another in the formation of a chemical bond, that is, to predict its polarity. The very detailed results given previously for the charge distributions of the diatomic hydrides are not generally available and there is a need for an empirical method which will allow us to estimate the polarity of any chemical bond. It is possible to define for an element a property known as its *electronegativity*, which provides a qualitative estimate of the degree of polarity of a bond. Electronegativity is defined as the ability of an atom in a molecule to attract electrons to itself. The concept of an electronegativity scale for the elements was proposed by Pauling.

The electron affinity of an atom provides a direct measure of the ability of an atom to attract and bind an electron:

$$X + e^- \rightarrow X^-, \text{where } \Delta E = A_X$$

where A_X denotes the electron affinity of atom X. For the reactions of two elements, X and Y, with free electrons, the relative values of the electron affinities A_X and A_Y provide a measure of the relative independent tendencies of X and Y to change into X^- and Y^-. However, we are interested in the reaction of X with Y and in being able to predict whether the X—Y bond will be polar in the sense X^+Y^- or X^-Y^+. The electron which is to be partially or wholly gained by X or Y is not a free electron but is bound to the atom Y or X respectively. Consequently we are interested in the relative energies of the following two processes:

$$X + Y \rightarrow X^+ + Y^-, \; \Delta E_1 = I_X + A_Y$$

$$X + Y \rightarrow X^- + Y^+, \; \Delta E_2 = I_Y + A_X$$

For reaction to be favoured over reaction, not only must Y have a high electron affinity, it is also necessary that X have a low ionization potential. We would expect the bonding electrons to be approximately equally shared in the X—Y bond, if $\Delta E_1 = \Delta E_2$, as neither extreme structure is favoured over the other. Thus the condition for a non-polar covalent bond is

$$I_X + A_Y \;=\; I_Y + A_X$$

or, collecting quantities for a given atom on one side of the equation,

$$I_X - A_X \;=\; I_Y + A_Y$$

Equation states that a non-polar bond will result when the difference between the ionization potential and the electron affinity is the same for both atoms joined by the bond. If the quantity I_X - A_X is greater than $I_Y - A_Y$, then the product X^-Y^+ will be energetically favoured over X^+Y^-. Thus the quantity $(I - A)$ provides a measure of the ability of an atom to attract electrons (or electronic charge density) to itself *relative* to some other atom. The electronegativity, denoted by the symbol χ, is defined to be proportional to this quantity:

$$\aleph_X \propto \left(I_X - A_X\right)$$

The electronegativities of the elements in the first few rows of the periodic table are given in Table.

Table. Some Electronegativity Values

H						
2.1						
Li	Be	B	C	N	O	P
1.0	1.5	2.0	2.5	3.0	6.5	4.0
Na	Mg	Al	Si	P	S	Cl
0.9	1.2	1.5	1.8	2.1	2.5	3.0
K	Ca					Br
0.8	1.0					2.8

As expected, the electronegativity increases from left to right across a given row of the periodic table and decreases down a given column. The greater the difference in the electronegativity values for two atoms, the greater should be the disparity in the extent to which the bond density is shared between the two atoms. Pauling has given empirical expressions which relate the electronegativity difference between two elements to the dipole moment and to the strength of the bond.

INTERACTION BETWEEN MOLECULES

The properties observed for matter on the macroscopic level are determined by the properties of the constituent molecules and the interactions between them. The polar or non-polar character of a molecule will clearly be important in determining the nature of its interactions with other molecules. There will be relatively strong forces of attraction acting between molecules with large dipole moments. To a first approximation, the energy of interaction between dipolar molecules can be considered as completely electrostatic in origin, the negative end of one molecule attracting the positive end of another.

The presence of intermolecular forces accounts for the existence of solids and liquids. A molecule in a condensed phase is in a region of low potential energy, a potential well, as a result of the attractive forces which the neighbouring molecules exert on it. By supplying energy in the form of heat, a molecule in a solid or liquid phase can acquire sufficient *kinetic energy* to overcome the *potential energy* of attraction and escape into the vapour phase. The vapour pressure (the pressure of the vapour in equilibrium with a solid or liquid at a given temperature) provides a measure of the tendency of a molecule in a condensed phase to escape into the vapour; the larger the vapour pressure, the greater the escaping tendency. The average kinetic energy of the molecule in the vapour is directly proportional to the absolute temperature. Thus the observation of a large vapour pressure at a low temperature implies that relatively little kinetic energy is required to overcome the potential interactions between the molecules in the condensed phase.

The only potential interactions possible between non-polar, covalently bonded molecules are of the van der Waals' type as previously discussed for the interaction between two helium atoms at large internuclear separations. Molecules such as H_2 and N_2 have closed shell electronic structures in the same sense that helium does; all of the valence electrons are paired and no further chemical bonding may occur. The small polarizations of the charge densities induced by the long-range interactions of closed shell atoms or molecules result in only weak forces of attraction. The low boiling points (the temperature at which the vapour pressure above the liquid phase equals one atmosphere) observed for substances composed of molecules which can interact only through a van der Waals' type force are, therefore, understandable. Table lists the normal boiling points for a number of representative compounds.

Table. Normal Boiling Points (°K)

Substance	BP	Substance	BP	Substance	BP
He	4.2	NH_3	240	NaCl	1686
H_2	20.4	HF	292	LiF	1949
N_2	77.4	H_2O	373	BeO	4100
Ar	87.4				

An argon atom is larger than a helium atom and its outer charge density is not bound as tightly as that in helium. Consequently, the charge density of argon is more polarizable than that of helium and the forces of attraction between argon atoms and hence its normal boiling point are correspondingly greater. These same forces do, of course, operate in the gas phase as well and are the cause of the observed deviations from ideal gas behaviour.

The interactions between polar molecules such as HF and H_2O will be much larger and their normal boiling points greater than those observed for the non-polar molecules. When hydrogen is present at the positive end of a polar bond, the dipolar interactions are particularly strong and are given a special name, hydrogen-bonded interactions. The hydrogen bond increases in strength as the electronegativity of the atom to which the H is chemically bonded increases. Liquid hydrogen fluoride consists of chains of molecules joined end to end; each hydrogen of one molecule is attracted to the fluorine of the next. In liquid water, each water molecule is hydrogen bonded to four other water molecules. This accounts for what appears to be an anomalously high boiling point for water when compared with the values observed for the neighbouring hydride molecules NH_3 and HF.

The condensed phases so far considered are called molecular solids or molecular liquids because the identity of the individual molecule is largely retained. As the forces between the molecules become larger, the point of view of regarding a solid as a collection of individual, interacting molecules becomes less satisfactory. In the limiting case of the strong interactions which exist between the ions in an ionic crystal, the concept of a discrete molecule in the solid phase ceases to exist. In solid KCl, for example, the potassium and chloride ions exist as separate entities; each potassium ion is in contact with six chloride ions, which in turn are each in contact with six potassium ions. Each ion attracts its six neighbouring ions equally and thus the structure is symmetrical and therefore cubic; six ions of one sign occupy the centres of the faces of a regular cube with an ion of opposite sign at its centre.

The number of nearest neighbours a given ion has in an ionic crystal is determined by the relative sizes of the positive and negative species. The Be^{+2} ion is considerably smaller than O^{-2} and the basic structure of BeO is tetrahedral, each ion surrounded by four ions of opposite charge. The strong electrostatic forces between the ions in a crystal are reflected in the high boiling points recorded in Table for the ionic compounds.

A QUANTITATIVE DEFINITION OF ATOMIC CHARGE

A theory of atoms in molecules (AIM) has been developed that enables one to define an atom in a molecule and all of its properties. The theoretical definition of an atom is based on the properties of the experimentally observable charge density. This theory is described elsewhere in this web site, but we present here a brief introduction to enable a quantitative definition of the charge on atom in a molecule.

A molecule, or a crystal, is partitioned into individual atoms in terms of surfaces that satisfy a particular condition on the electron density. The partitioning is a direct result of the density exhibiting a maximum at the position of each nucleus. As a consequence, the density passes through some minimum value before again attaining some maximum value at a neighbouring nucleus. This fundamental property of a charge distribution is illustrated in the accompanying figure, Fig which displays the electron density in a plane of the nuclei for the *LiF* molecule. The distribution is dominated by the two nuclear maxima. The point of minimum density along the line linking the two nuclei – the saddle point in the density – is a starting point for two paths of steepest descent away from this point, paths that define the boundary separating the atoms in this plane. The collection of such paths for all planes obtained by rotation about the axis defines the interatomic surface. The interatomic surfaces defined in this manner yield a 'natural partitioning' of a molecule into atoms, as illustrated in Fig. *A1* for *LiF* and in Figures for the second- and third-row diatomic hydrides.

The properties of the atoms defined by these 'natural' surfaces are determined by quantum mechanics. Thus, not only does quantum mechanics predict the properties of a molecule, it also predicts the properties of its individual constituent atoms. The theory of atoms in molecules (*AIM*) equates every property of a molecule to the sum of its atomic contributions. Among the properties of immediate interest is the charge on an atom. The electron population of an atom A - its average number of electrons - a quantity denoted by $N(A)$, is obtained by integrating the electron density over the space of the atom up to its atomic boundaries, as given in equation where the electron density at a point denoted by the position vector r is denoted by $\rho(r)$

$$N(A) = \int_A \rho(r)\,dr$$

The subscript A on the integral means that the integration is carried out over the space of the atom A up to its atomic boundary, that is, its interatomic surface. The charge on the atom, $q(A)$, is the difference between the nuclear charge Z_A and its electron population, as given in equation $q(A) = Z_A - N(A)$

Another useful property determined by the electron density is the atomic polarization. In a free atom the centroid of negative charge coincides with the nucleus and there is no dipole moment, that is, no separation of charge. When an atom is in chemical combination, its density becomes polarized, either towards or away from a neighbouring atom. One can easily determine the displacement of the centroid of negative charge from the position of the nucleus by determining the average (vector) distance of every element of the electron density from the nucleus of the atom. This corresponds to integrating the product of the density at each position r, with the vector distance of this point from the nucleus, a distance denoted by the vector r_A to obtain the atomic moment $M(A)$.

$$M(A) = \int_A r_A \rho(r)\,dr$$

The atomic moment is a vector because it has a direction as well as a magnitude.

The dipole moment μ of a diatomic molecule A-B, when expressed in terms of atomic contributions, is given by a charge transfer term, equal to the charge on atom B multiplied by R, the vector distance from the nucleus of A and to that of B, and a polarization term given by the sum of the two atomic moments. The dipole moment of A-B is therefore given as in equation.

$$\mu = q(B)R + M(A) + M(B)$$

where all three terms are directed along the internuclear axis. The sign of the charge transfer term $q(B)R$ is determined by the sign of $q(B) = -q(A)$. The model expression for μ given for *LiF* in the text assumes the complete transfer of one electron from *Li* to *F* and equates the dipole moment to the charge transfer term alone, that is, to the first term of equation A-4. We can use the molecular density distribution for LiF displayed in Fig. to calculate the charges on the atoms and determine the atomic polarizations to illustrate how the atomic contributions determine the molecular dipole. The charge on *F* is $q(F) = -0.938e$, close to the value of $-1e$ assumed in the model calculation. As already discussed, the atoms are in general polarized in a direction counter to the direction of the charge transfer. In *LiF* the charge transfer is from $Li \rightarrow F$ and hence the fluorine atom is polarized towards the lithium atom which in turn is polarized away from the fluorine atom. The length of the vector *R* in *LiF* is 1.564 Å and expressing the charge on the fluorine atom in esu, the charge transfer contribution to the dipole moment is – 7.05 debyes. The atomic polarizations are $M(F) = +0.714$ and $M(Li) = +0.043$ debyes to yield a dipole moment with a magnitude of 6.29 debyes, the experimental value being 6.28 debyes.

Properties Determined by the Electronic Charge Density

Pauling defined electronegativity to be 'the power of an atom in a molecule to attract electrons to itself.' This concept has proved to be extremely useful and it is reflected in the net charges on the atoms found in diatomic molecules, as determined by the theory of atoms in molecules. The atomic charges in a diatomic molecule are a direct measure of the relative abilities of the two atoms to attract and bind electronic charge within their basins. The variation in the charge on atom *A* in *AB* where A and B vary across the second row of the periodic table, $Li \rightarrow F$ and including hydrogen.

Each atom withdraws charge from elements to the left of it and donates charge to those on its right, with *H* appearing between *C* and *N*. The orderings are as anticipated with *C* and *H* possessing almost equal electronegativities. The electronegativity of *C* relative to *H* increases with the degree of unsaturation and with the extent of geometric strain. This result is anticipated on the basis of the orbital model which predicts the electronegativity of C to increase as the s character of its hybrid bonds to H increases. Most of the secondary variation in charges across the table are explicable in terms of the extent of charge transfer being limited by either the number of valence electrons on the donor or vacancies on the acceptor. The charges on the third-row elements $Na \rightarrow Cl$ are also given relative to H in their hydrides and, as anticipated, H advances towards the electronegative end of the scale in this row relative to its position *vis-à-vis* the second-row elements. Unexpectedly, sodium and magnesium are slightly less electropositive than their second-row congeners.

The extent and direction of charge transfer and its effect on the charge distribution are reflected in the behaviour of the interatomic surface which is indicated for each molecule. In LiH the surface envelops what is essentially a Li ion while in HF the total charge distribution is dominated by the forces exerted by the F nucleus. The charges $q(H)$ of AH are also characteristic of the stable polyatomic species AH_n the two values usually differing by less than 0.05 e, and they reflect the chemical behaviour of the hydrides. The hydrides of Li, Be, and B, for example, are all hydridic, expelling molecular hydrogen from water and, for all of these, $q(H) < 0$. There is a sharp break in the value of $q(H)$ for methane for which $q(H) \approx 0$ and this is a faithful reflection of the non-polar nature of this molecule. It has a low solubility in water and does not dissociate. The remaining hydrides, NH_3, H_2O, and HF, are all increasingly polar with $q(H) > 0$ and the ordering of the charges accounts for the aqueous solution of ammonia being basic and that of HF being acidic.

The ability to determine the charge on an atom in a molecule removes the necessity of defining a numerical electronegativity scale. The concept, however, remains useful and one may use the atomic populations to demonstrate that they recover the basic idea underlying electronegativity—to predict the degree of charge transfer between two atoms. Since hydrogen can either donate or accept but a single electron, the electron population of hydrogen in AH may be used to define an electronegativity per electron of A relative to hydrogen. This electronegativity is given by $X(A) = 1-N(H)_{AH}$, where $N(H)_{AH}$ is the population of H in AH. A positive or negative value for $X(A)$ implies that A has a greater or lesser bonding electron affinity than does hydrogen, respectively. If the $X(A)$ are meaningful, then the difference $|X(A) - X(B)|$ should determine the charge transfer per valence electron in AB. Using this concept, the population of A in AB is predicted to be

$$N(A)_{AB} = \left[N(A)_a - \{X(B) - X(A)\}v\right]$$

where v is the number of valence electrons on the donor A or the number of vacancies on the acceptor B, whichever value is limiting. $N(A)_a$ is the electron population of the isolated A atom. Examples of predicted and actual atomic populations for A are: *NF*, 6.56 (6.56); *NO*, 6.46 (6.50); *CF*, 5.21 (5.22); *CO*, 4.78 (4.65); *CN*, 4.94 (4.88); *LiC*, 2.12 (2.12). With the added stipulation that the charge transferred per vacancy, $|X(A) - X(B)|$, cannot exceed v, then all diatomic fluoride populations are predicted to within a maximum error of 0.08 electrons. This method of predicting the charge transfer between atoms yields significantly larger errors only for some compounds of the elements *Be* and *B*.

The charge distributions of the diatomic hydrides illustrate a general phenomenon—that a significant degree of interatomic charge transfer is accompanied by a polarization of the valence densities of the atoms in a direction counter to that of the charge transfer. The polarizations are in response to the electric field created by the charge transfer, the acceptor atom polarizing towards the positively charged donor atom which is itself polarized away from the negatively charged acceptor. This polarization of the donor atom is particularly pronounced when it possesses a greater number of valence electrons than there are vacancies on the acceptor atom, as illustrated by the data for the diatomic hydrides given in Table. This Table lists the atomic quantities that determine the molecular dipole moment, as given in equation; the charges on the atoms, the atomic polarizations, the charge transfer contribution to the molecular dipole moment under the heading $\mu(CT) = q(H)R = -q(A)R$, the dipole moment for each molecule $\mu(AH)$ and the nonbonded radius of the A atom, $r_n(A)$.

In general, the magnitude of the molecular dipole is less than that $\mu(CT)$, the charge transfer contribution, because of the opposing atomic polarizations. The polarizations of *Li* and *Na* in their hydrides are quite small as they correspond to tightly bound core densities. However, in some instances such as *BH*, the atomic polarizations determine the direction of the molecular moment. For the second row, the atomic polarizations are largest for the diffuse valence density on *Be* and *B*. They are larger still for their third-row congeners and *Si* where, because of the larger, 10-electron *K-L* core, the valence density is less tightly bound and more polarizable.

Attempts to assign atomic charges on the basis of measured dipole moments are unrealistic as such a procedure ignores the polarizations of the atomic densities. Such an attempt corresponds to assuming the molecular charge distribution to be composed of a set of spherically symmetric atomic densities, each centred on its own nucleus, a physically unacceptable model even in the limit of an ionic system. It should be evident from a comparison of the charge distribution in the non-bonded region of the *A* atoms that the reduction in magnitude or reversal in sign of the dipole moment, which occurs after *LiH* in the second-row and after *Na* in the third-row, is a consequence of an atomic polarization and is not indicative of a sudden increase in the electronegativity of the A atom. The extent of the physical distortion of those atoms for which the atomic polarizations are greatest is reflected in the values of their non-bonded radii. A *non-bonded radius*, $r_n(\Omega)$, is defined as the axial distance from a nucleus to an outer contour of the charge density on its non-bonded side. The 0.001aû contour is chosen since the corresponding density envelopes provide good approximations to the experimentally determined van der Waals sizes and corresponding radii for molecules in the gas phase. The non-bonded radii for *Li* and *Na* are close to the values for the corresponding singly-charged ions while those for the strongly back-polarized atoms are all considerably greater than are their values in the free atomic state.

The presence of such a large and diffuse (weakly bound) charge distribution has important chemical consequences, imparting to the molecule the characteristics of a strong Lewis base. A classic example of this behaviour is the carbon atom in the *CO* molecule. This molecule has a near-zero dipole moment because of very pronounced polarizations of the atomic densities, particularly that of carbon, which oppose the considerable charge transfer moment. The charge on oxygen is –1.33 e and the magnitudes of the opposing atomic dipoles are $|M(O)| = 0.98$ au and $|M(C)| = 1.72$ au with the non-bonded radius on carbon exceeding its free atomic value by 0.15 au. The physical importance of the atomic polarization of carbon is reflected in the ability of *CO* to act as a Lewis base, particularly in the formation of metal carbonyls. The considerable difference in the electronegativities of *C* and *O* is reflected in the relatively large dipole moment, $|\mu| = 1.11$ au, of the formaldehyde molecule, $H_2C{=}O$. The charge transfer from *C* to *O* in formaldehyde where $q(O) = -\ 1.24\ e$, is only slightly less than it is in *CO*. Unlike *CO*, however, the charge transfer contribution dominates the final moment in formaldehyde because of the close to halving of the atomic dipole on carbon which results from the use of its non-bonded density in the formation of non-polar bonds to the hydrogen atoms.

Chapter 6

Molecular Orbital

There is a second major theory of chemical bonding whose basic ideas are distinct from those employed in valence bond theory. This alternative approach to the study of the electronic structure of molecules is called molecular orbital theory. The theory applies the orbital concept, which was found to provide the key to the understanding of the electronic structure of atoms, to molecular systems.

The concept of an orbital, whether it is applied to the study of electrons in atoms or molecules, reduces a many-body problem to the same number of one-body problems. In essence an orbital is the quantum mechanical description (wave function) of the motion of a single electron moving in the average potential field of the nuclei and of the other electrons which are present in the system. An orbital theory is an approximation because it replaces the instantaneous repulsions between the electrons by some average value. The difficulty in obtaining an accurate description of an orbital is the difficulty in determining the average potential field of the other electrons. For example, the 2*s* orbital in the lithium atom is a function which determines the motion of an electron in the potential field of the nucleus and in the average field of the two electrons in the 1*s* orbital. However, the 1*s* orbital is itself determined by the nuclear potential field and by the average potential field exerted by the electron in the 2*s* orbital. Each orbital is dependent upon and determined by all the other orbitals of the system. To know the form of one orbital we must know the forms of all of them. This problem has a mathematical solution; the exploitation of this solution has proved to be one of the most powerful and widely used methods to obtain information on the electronic structure of matter.

A molecular orbital differs from the atomic case only in that the orbital must describe the motion of an electron in the field of more than one nucleus, as well as in the average field of the other electrons. A molecular orbital will in general, therefore, encompass all the nuclei in the molecule, rather than being centred on a single nucleus as in the atomic case. Once the forms and properties of the molecular orbitals are known, the electronic configuration and properties of the molecule are again determined by assigning electrons to the molecular orbitals in the order of increasing energy and in accordance with the Pauli exclusion principle.

In valence bond theory, a single electron pair bond between two atoms is described in terms of the overlap of atomic orbitals which are centred on the nuclei joined by the bond. In molecular orbital theory the bond is described in terms of a single orbital which is determined by the field of both nuclei. The two theories provide only a first approximation to the chemical bond.

We shall begin our discussion of molecular orbital theory by applying the theory to the discussion of the bonding in the homonuclear diatomic molecules.

ANGULAR MOMENTUM IN DIATOMIC MOLECULES

The spatial symmetries of atomic orbitals and the number of each symmetry type are determined by the angular momentum of the electron. The orbitals are in fact labelled by the angular momentum quantum numbers, l and m, which along with the principal quantum number n, completely specify the orbital. Angular momentum plays a similar role in determining the symmetries and number of orbitals of each symmetry species in the molecular case.

In an atom all of the angular momentum is electronic in origin. In the molecular case, the molecule as a whole rotates in space and the nuclei contribute to the total angular momentum of the system.

The nuclei and the electrons of a diatomic molecule can rotate about both of the axes which are perpendicular to the bond axis.

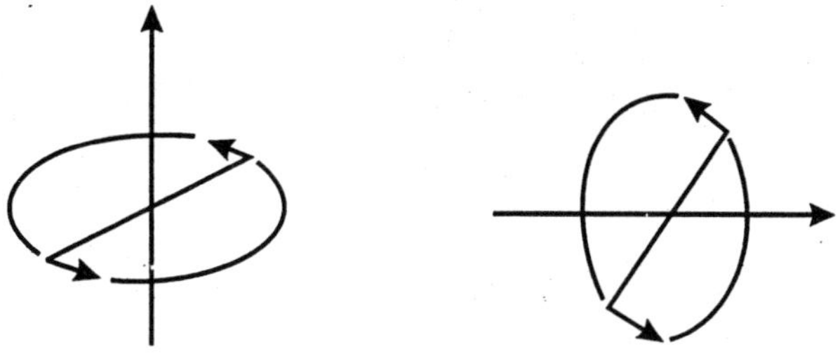

Fig. Two Rotational Axis for a Diatomic Molecule.

In a classical analogue the electrons and the nuclei exchange angular momentum during these rotations and the angular momentum of the electrons is not separately conserved. Thus the magnitude of the total electronic angular momentum in a diatomic molecule, unlike the atomic case, is not quantized. Instead, the magnitude of the total angular momentum, nuclei and electrons, is quantized. Only the electrons, however, may rotate *about* the internuclear axis and this component of the angular momentum is entirely electronic in origin. As long as the molecule is left undisturbed, this one component of the angular momentum remains fixed in value and its magnitude, is, therefore, quantized.

The angular momentum *vector* for rotation about the bond will lie along the bond axis. This vector represents the component of the total angular momentum vector along the internuclear axis. As in the atomic case, quantum mechanics restricts the values of the component of the total angular momentum vector along a given axis to integral multiples of $(h/2\pi)$. The quantum number in this case is denoted by the Greek letter λ (lambda). It is analogous to the quantum number m in the atomic case. The possible values for λ are

$$\lambda = 0,1,2,3,...$$

Since the rotation may occur in the clockwise or anticlockwise sense about the axis, the angular momentum vector component may be pointed in either direction along the bond.

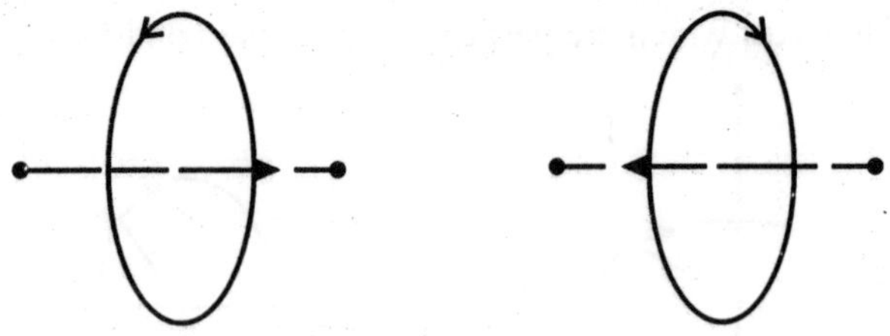

Fig. The two directions for the orbital angular momentum vector λ for the rotation of an electron about the internuclear axis of a diatomic molecule.

Correspondingly, the allowed values of the angular momentum about the internuclear axis are 0, ±1 $(h/2\pi)$, $\pm 2(h/2\pi)$, etc., or in general, $\pm\lambda(h/2\pi)$. Thus when λ is different from zero, each energy level is doubly degenerate corresponding to the two possible directions for the component λ along the bond axis.

The molecular orbitals are labelled according to the values of the quantum number λ. When λ = 0, they are called σ orbitals; when λ = 1, π orbitals; when λ = 2, δ orbitals, etc. This is analogous to the labelling of the atomic orbitals as *s*, *p*, *d*, . . . , as determined by their *l* value.

We know less about the angular momentum of an electron in a diatomic molecule than in an atom. In the atomic case it is possible to determine the magnitude of the total angular momentum, as given by the quantum number *l*, and the magnitude of one of its components, as given by the quantum number *m*. In a linear molecule our knowledge is more restricted and we are limited to a single quantum number λ, which determines only the component of angular momentum about the bond axis.

SYMMETRY CONSIDERATIONS

The potential field of a nucleus in an atom is spherically symmetric, depending only on the distance between the nucleus and the electron. Consequently the spatial symmetries of atomic orbitals are completely determined by the angular momentum quantum numbers *l* and *m*. When spherical polar coordinates rather than cartesian coordinates are used to describe the orbitals the dependence of the orbitals on the angles θ and ϕ is determined by their angular momentum quantum numbers.

Only the radial dependence (the dependence of the orbital on the coordinate r, the distance between the nucleus and the electron) differs between orbitals with the same l and m values but different values of n.

The potential field of the nuclei in a linear molecule possesses cylindrical symmetry. In terms of a cylindrical coordinate system the single angular momentum quantum number λ determines the dependence of the molecular orbitals on the angle ϕ, a dependence determining the symmetry of the orbital for a rotation about the internuclear axis. The dependence of the molecular orbitals on ρ and z is left undetermined.

The forms of the orbitals are not as fully determined by the angular momentum quantum numbers in a molecule as in an atom. However, we may further characterize and label the orbitals for a molecular system by taking advantage of the symmetry possessed by the molecule. The symmetry of the potential field in which an electron moves places very severe restrictions on the possible forms of the orbitals. This is a very general and powerful result. Indeed, the angular dependence of orbitals and wave functions and their angular momentum quantum numbers may be completely determined solely by a consideration of the rotational symmetry of a system.

We may illustrate the role which symmetry plays in determining the form of an orbital by considering the symmetry properties of the orbitals obtained in Section 2 for the case of an electron restrained to move on a line of fixed length. Let us shift the origin of the x-axis in the plots of the orbitals to the mid-point of the line, thereby changing the values of the coordinates of the two end points from 0 and L to $-L/2$ and $+L/2$ respectively. Next let us denote by the symbol R the operation of reflection through the origin, an operation which replaces each value of x by $-x$. For example, the end points $x = -L/2$ and $x = +L/2$ are interchanged by the reflection operator R.

The first point to note about the operation of reflection is that its application leaves the phyiscal system itself unchanged. The potential in which the electron moves is assumed to be of constant value along the x-axis. The reflection operator simply interchanges the two halves of the line leaving the system unchanged. The potential is said to be *invariant* to the operation of reflection through the origin.

What is the effect of R on the wave functions or orbitals? When R operates on $\psi_1(x)$ (that is, when $\psi_1(x)$ is reflected through the origin) the result is obviously to change $\psi_1(x)$ into itself:

$$
\begin{aligned}
R\psi_1(x) &= \psi_1(-x) \\
&= (+1)\psi_1(x)
\end{aligned}
$$

The reflected function $\psi_1(-x)$ is indistinguishable from $\psi_1(x)$.

The result of operating on $\psi_1(x)$ with the operator R is to leave the function *unchanged*. $\psi_1(x)$ is said to be *symmetric* with respect to a reflection through the origin. The operation of R on $\psi_2(x)$ yields a different result:

$$
\begin{aligned}
R\psi_2(x) &= \psi_2(-x) \\
&= (+1)\psi_2(x)
\end{aligned}
$$

It is obvious from Fig that the reflection of $\psi_2(x)$ through the mid-point *changes its sign*, the reflected function $\psi_2(-x)$ is the negative of the unreflected function $\psi_2(x)$. Such a function is said to be *antisymmetric* with respect to a reflection at the origin. Every orbital for this system is either symmetric (those with odd n values) or antisymmetric (those with even n values) with respect to the symmetry operation of reflection.

Any orbital which was neither symmetric nor antisymmetric but was instead simply unsymmetrical with respect to reflection would when squared yield an unsymmetrical probability distribution. An unsymmetrical probability distribution implies that the electron is more likely to be found on one half of the x-axis than on the other. This is a phy sically unacceptable result since there are no forces acting on the electron which would favour one end of the line over the other. Only orbitals which are either symmetric or antisymmetric yield density distributions which properly reflect the symmetry of the system, that is, density distributions which are themselves symmetrical with respect to reflection at the mid-point of the line.

Thus we conclude that the only wave functions resulting in physically acceptable probability distributions are those which are either symmetrical or antisymmetrical with respect to any symmetry operation which changes the physical system into itself. This statement is always true for non-degenerate wave functions, but must be amended somewhat for the action of some symmetry operations on a degenerate set of wave functions.

We shall use only one of the many symmetry elements possessed by a homonuclear diatomic molecule to further characterize and classify the molecular orbitals. A homonuclear diatomic molecule possesses a centre of symmetry and the corresponding operator is called the inversion operator. The action of this operator, denoted by the symbol i, is to replace the x, y, z coordinates of every point in space by their negatives $-x$, $-y$, $-z$. This corresponds to an inversion (or reflection) of every point through the origin or centre of symmetry of the molecule.

The action of the inversion operator on the nuclear coordinates simply interchanges one nucleus for the other. Since the nuclei possess identical charges, the nuclear framework is left unchanged and the potential exerted by the nuclei is invariant to the operation of inversion. Thus every molecular orbital for a homonuclear molecule must be either symmetric or antisymmetric with respect to the inversion operator. Orbitals which are left unchanged by the operation of inversion (are symmetric) are labelled with a subscript g, while those which undergo a change in sign (are antisymmetric) are labelled u. The symbols g and u come from the German words "gerade" and "ungerade" meaning "even" and "odd" respectively.

MOLECULAR ORBITALS FOR HOMONUCLEAR DIATOMICS

While the specific forms of the molecular orbitals (their dependence on ρ and z in a cylindrical coordinate system) are different for each molecule, their dependence on the angle ϕ as denoted by the quantum number λ and their *g* or *u* behaviour with respect to inversion are completely determined by the symmetry of the system. These properties are common to all of the molecular orbitals for homonuclear diatomic molecules. In addition, the relative ordering of the orbital energies is the same for nearly all of the homonuclear diatomic molecules. Thus we may construct a molecular orbital energy level diagram, similar to the one used to build up the electronic configurations of the atoms in the periodic table. The molecular orbital energy level diagram is as fundamental to the understanding of the electronic structure of diatomic molecules as the corresponding atomic orbital diagram is to the understanding of atoms.

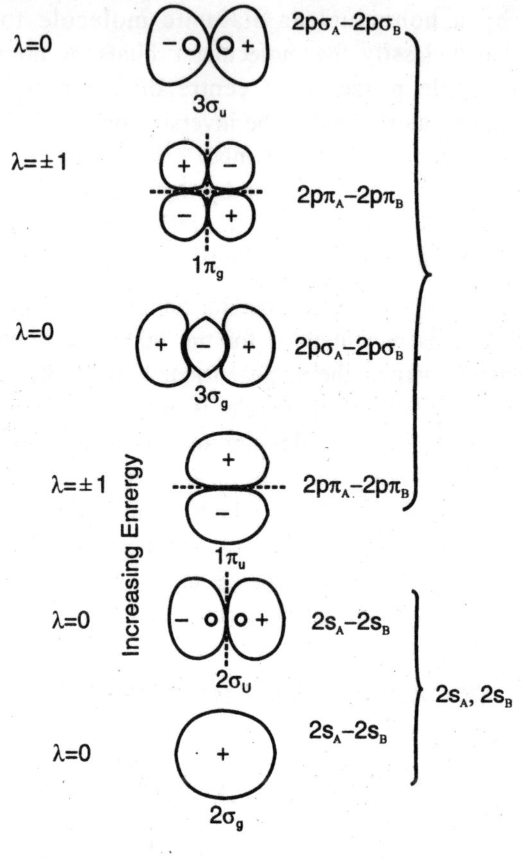

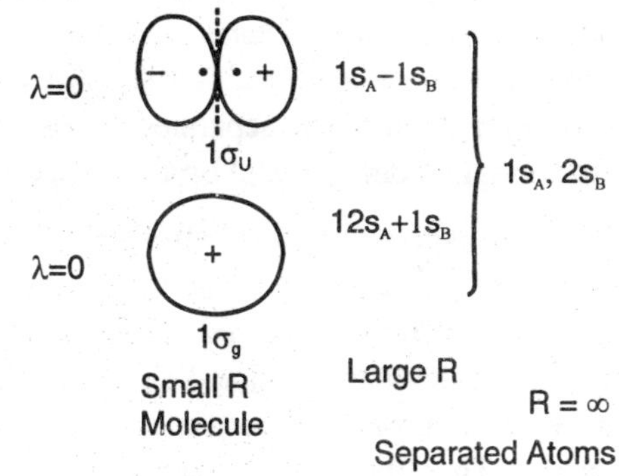

Fig. Molecular Orbital Energy level Diagram for Homonuclear Diatomic Molecules

Molecular orbitals exhibit the same general properties as atomic orbitals, including a nodal structure. The nodal properties of the orbitals are indicated in Fig. Notice that the nodal properties correctly reflect the *g* and *u* character of the orbitals. Inversion of a *g* orbital interchanges regions of like sign and the orbital is left unchanged. Inversion of a *u* orbital interchanges the positive regions with the negative regions and the orbital is changed in sign.

An orbital of a particular symmetry may appear more than once. When this occurs a number is added as a prefix to the symbol. Thus there are $1\sigma_g$, $2\sigma_g$, $3\sigma_g$, etc. molecular orbitals just as there are 1*s*, 2*s*, 3*s*, etc. atomic orbitals. The numerical prefix is similar to the principal quantum number *n* in the atomic case. As *n* increases through a given symmetry set, for example, $1\sigma_g$, $2\sigma_g$, $3\sigma_g$, the orbital energy increases, the orbital increases in size and consequently concentrates charge density further from the nuclei, and finally the number of nodes increases as *n* increases. All these properties are common to atomic orbitals as well.

We may obtain a qualitative understanding of the molecular orbital energy level diagram by considering the behaviour of the orbitals under certain limiting conditions. The molecular orbital must describe the motion of the electron for all values of the internuclear separation; from $R = \infty$ for the separated atoms, through $R = R_e$, the equilibrium state of the molecule, to $R = 0$, the united atom obtained when the two nuclei in the molecule coalesce (in a hypothetical reaction) to give a single nucleus. Hence a molecular orbital must undergo a continuous change in form. At the limit of large R it must reduce to some combination of atomic orbitals giving the proper orbital description of the separated atoms and for $R = 0$ it must reduce to a single atomic orbital on the united nucleus.

Consider, for example, the limiting behaviour of the $1\sigma_g$ orbital in the case of the hydrogen molecule. The most stable state of H_2 is obtained when both electrons are placed in this orbital with paired spins giving the electronic configuration $1\sigma_g^{\ 2}$. For large values of the internuclear separation, the hydrogen molecule dissociates into two hydrogen atoms. Thus the limiting form of the $1\sigma_g$ molecular orbital for an infinite separation between the nuclei should be a sum of $1s$ orbitals, one centred on each of the nuclei. If we label the two nuclei as A and B we can express the limiting form of the $1\sigma_g$ orbital as

$$1\sigma_g \rightarrow \left(1s_A + 1s_B\right)$$

where $1s_A$ is a $1s$ orbital centred on nucleus A, and $1s_B$ is a $1s$ orbital centred on nucleus B. This form for the $1\sigma_g$ orbital predicts the correct density distribution for the system at large values of R. Squaring the function $(1s_A + 1s_B)$ we obtain for the density

$$\left(1s_A \times 1s_A + 1s_B \times 1s_B + 2\left(1s_A \times 1s_B\right)\right)$$

The first two terms denote that one electron is on atom A and one on atom B, both with 1s atomic density distributions. The cross term $2 \times 1s_A \times 1s_B$ obtained in the product is zero since the distance between the two nuclei is so great that the overlap of the orbitals vanishes. Notice as well that the function $(1s_A + 1s_B)$ has the same symmetry properties as does the $1\sigma_g$ molecular orbital; it is symmetric with respect to both a rotation about the line joining the nuclei and to an inversion of the coordinates at the mid-point between the nuclei. The $1\sigma_g$ orbital for the *molecule* is said to *correlate* with the sum of $1s$ orbitals, one on each nucleus, for the *separated atom case*.

Consider next the limiting case of the separated atoms for the helium molecule. Of the four electrons present in He_2 two are placed in the $1\sigma_g$ orbital and the remaining two must, by the Pauli exclusion principle, be placed in the next vacant orbital of lowest energy, the $1\sigma_u$ orbital. The electronic configuration of He_2 is thus $1\sigma_g^2 1\sigma_u^2$. The $1\sigma_g$ orbital will correlate with the sum of the $1s$ orbitals for the separated helium atoms. Of the two electrons in the $1\sigma_g$ molecular orbital one will correlate with the $1s$ orbital on atom A and the other with the $1s$ orbital on atom B. Since each helium atom possesses two $1s$ electrons, the $1\sigma_u$ orbital must also correlate its electrons with $1s$ atomic functions on A and B. In addition, the correlated function in this case must be of u symmetry. A function with these properties is

$$1\sigma_u \rightarrow (1s_A - 1s_B)$$

The limiting density distribution obtained by squaring this function places one electron in a $1s$ atomic distribution on A, the other in a $1s$ atomic distribution on B. The sum of the limiting charge densities for the $1\sigma_g$ and $1\sigma_u$ molecular orbitals places two electrons in $1s$ atomic charge distributions on each atom, the proper description of two isolated helium atoms.

Every diatomic homonuclear molecular orbital may be correlated with either the sum (for σ_g and π_u orbitals) or the difference (for σ_u and π_g orbitals) of like orbitals on both separated atoms. By carrying out this correlation procedure for every orbital we may construct a molecular orbital correlation diagram which relates each of the orbital energy levels in the molecule with the correlated energy levels in the separated atoms. It is important to note that the symmetry of each orbital is preserved in the construction of this diagram. Consider, for example the molecular orbitals which correlate with the $2p$ atomic orbitals. The direction of approach of the two atoms defines a new axis of quantization for the atomic orbitals. The $2p$ orbital which lies along this axis is of σ symmetry while the remaining two $2p$ orbitals form a degenerate set of π symmetry with respect to this axis. The sum and difference of the $2p\sigma$ orbitals on each centre correlate with the $3\sigma_g$ and $3\sigma_u$ orbitals respectively, while the sum and the difference of the $2p\pi$ orbitals correlate with the π_u and π_g orbitals.

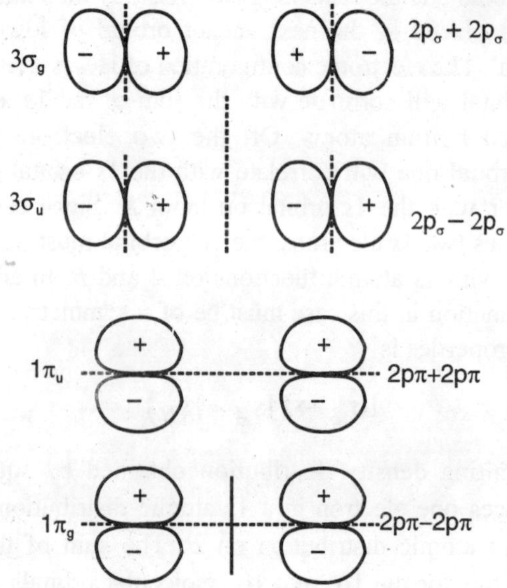

Fig. The correlated separated atom forms of the $3\sigma_g$, $3\sigma_u$, $1\pi_u$ and $1\pi_g$ molecular orbitals. The nodal planes are indicated by dashed lines. Only one component of each π orbital is shown.

For large values of the internuclear distance, each molecular orbital is thus represented by a sum or a difference of atomic orbitals centred on the two interacting atoms. As the atoms approach one another the orbitals on each atom are distorted by polarization and overlap effects. In general, the limiting correlated forms of the molecular orbitals are *not* suitable descriptions of the molecular orbitals for finite internuclear separations.

The correlation of the molecular orbitals with the appropriate atomic orbitals of the united atom is left as a problem for the reader.

We are now in a position to build up and determine the electronic configurations of the homonuclear diatomic molecules by adding electrons two at a time to the molecular orbitals with the spins of the electrons paired, always filling the orbitals of lowest energy first. We shall, at the same time, discuss the effectiveness of each orbital in binding the nuclei and make qualitative predictions regarding the stability of each molecular configuration.

Hydrogen. The two electrons in the hydrogen molecule may both be accommodated in the $1\sigma_g$ orbital if their spins are paired and the molecular orbital configuration for H_2 is $1\sigma_g^2$. Since the $1\sigma_g$ orbital is the only occupied orbital in the ground state of H_2, the density distribution shown previously in Fig for H_2 is also the density distribution for the $1\sigma_g$ orbital when occupied by two electrons. The remarks made previously regarding the binding of the nuclei in H_2 by the molecular charge distribution apply directly to the properties of the $1\sigma_g$ charge density. Because it concentrates charge in the binding region and exerts an attractive force on the nuclei the $1\sigma_g$ orbital is classified as a *bonding orbital.*

Excited electronic configurations for molecules may be described and predicted with the same ease within the framework of molecular orbital theory as are the excited configurations of atoms in the corresponding atomic orbital theory. For example, an electron in H_2 may be excited to any of the vacant orbitals of higher energy indicated in the energy level diagram. The excited molecule may return to its ground configuration with the emission of a photon. The energy of the photon will be given approximately by the difference in the energies of the excited orbital and the $1\sigma_g$ ground state orbital. Thus molecules as well as atoms will exhibit a line spectrum. The electronic line spectrum obtained from a molecule is, however, complicated by the appearance of many accompanying side bands. These have their origin in changes in the vibrational energy of the molecule which accompany the change in electronic energy.

Helium. The electronic configuration of He_2 is $1\sigma_g^2\ 1\sigma_u^2$. A σ_u orbital, unlike a σ_g orbital, possesses a node in the plane midway between the nuclei and perpendicular to the bond axis. The $1\sigma_u$ orbital and all σ_u orbitals in general, because of this nodal property, cannot concentrate charge density in the binding region. It is instead concentrated in the antibinding region behind each nucleus.

The σ_u orbitals are therefore classified as *antibonding*. It is evident from the form of density distribution for the $1\sigma_u$ orbital that the charge density in this orbital pulls the nuclei apart rather than drawing them together. Generally, the occupation of an equal number of σ_g and σ_u orbitals results in an unstable molecule. The attractive force exerted on the nuclei by the charge density in the σ_g orbitals is not sufficient to balance both the nuclear force of repulsion and the antibinding force exerted by the density in the σ_u orbitals. Thus molecular orbital theory ascribes the instability of He_2 to the equal occupation of bonding and antibonding orbitals. Notice that the Pauli exclusion principle is still the basic cause of the instability. If it were not for the Pauli principle, all four electrons could occupy a σ_g-type orbital and concentrate their charge density in the region of low potential energy between the nuclei. It is the Pauli principle, and not a question of energetics, which forces the occupation of the $1\sigma_u$ antibonding orbital.

The total molecular charge distribution is obtained by summing the individual molecular orbital densities for single or double occupation numbers as determined by the electronic configuration of the molecule. Thus the total charge distribution for He_2 is given by the sum of the $1\sigma_g$ and $1\sigma_u$ orbital densities for double occupation of both orbitals. The adverse effect which the nodal property of the $1\sigma_u$ orbital has on the stability of He_2 is very evident in the total charge distribution. Very little charge density is accumulated in the central portion of the binding region. The value of the charge density at the mid-point of the bond in He_2 is only 0.164 au compared to a value of 0.268 au for H_2.

We should reconsider in the light of molecular orbital theory the stability of He_2^+ and the instability of the hydrogen molecule with parallel spins, cases discussed previously in terms of valence bond theory. He_2^+ will have the configuration $1\sigma_g^2\,1\sigma_u^1$. Since the $1\sigma_u$ orbital is only singly occupied in He_2^+, less charge density is accumulated in the antibinding regions than is accumulated in these same regions in the neutral molecule. Thus the binding forces of the doubly-occupied $1\sigma_g$ density predominate and He_2 is stable. The electron configuration of H_2 is $1\sigma_g^1(\uparrow)1\sigma_u^1(\uparrow)$ when the electronic spins are parallel. The electrons must occupy separate orbitals because of the Pauli exclusion principle. With equal occupation of bonding and antibonding orbitals, the H_2 ($\uparrow\uparrow$)species is predicted to be unstable.

Lithium. The Li_2 molecule with the configuration $1\sigma_g^2 1\sigma_u^2 2\sigma_g^2$ marks the beginning of what can be called the second quantum shell in analogy with the atomic case. Since the $1\sigma_u$ antibonding orbital approximately cancels the binding obtained from the $1\sigma_g$ bonding orbital, the bonding in Li_2 can be described as arising from the single pair of electrons in the $2\sigma_g$ orbital. Valence bond theory, or a Lewis model for Li_2, also describes the bonding in Li_2 as resulting from a single electron pair bond. This is a general result. The number of bonds predicted in a simple Lewis structure is often found to equal the difference between the number of occupied bonding and antibonding orbitals of molecular orbital theory.

The forms of the orbital density distributions for Li_2 bear out the prediction that a single electron pair bond is responsible for the binding in this molecule.

The $1\sigma_g$ and $1\sigma_u$ density distributions are both strongly localized in the regions of the nuclei with spherical contours characteristic of $1s$ atomic distributions. The addition of just the doubly-occupied $1\sigma_g$ and $1\sigma_u$ orbital densities in Li_2 will yield a distribution which resembles very closely and may be identified with the doubly-occupied $1s$ or inner shell atomic densities on each lithium nucleus. Only the charge density of the pair of valence electrons in the $2\sigma_g$ orbital is delocalized over the whole of the molecule and accumulated to any extent in the binding region.

Thus there are cases where the molecular orbitals even at the equilibrium bond length resemble closely their limiting atomic forms. This occurs for inner shell molecular orbitals which correlate with the inner shell atomic orbitals on the separated atoms. Inner shell $1s$ electrons are bound very tightly to the nucleus as they experience almost the full nuclear charge and the effective radii of the $1s$ density distributions are less than the molecular bond lengths. Because of their tight binding and restricted extension in space, the inner electrons do not participate to any large extent in the binding of a molecule. Thus with the exception of H_2 and He_2 and their molecular ions, the $1\sigma_g$ and $1\sigma_u$ molecular orbitals degenerate into non-overlapping atomic-like orbitals centred on the two nuclei and are classed as nonbonding orbitals.

Beryllium. The configuration of Be_2 is $1\sigma_g^2 1\sigma_u^2 2\sigma_g^2 2\sigma_u^2$ and the molecule is predicted to form a weakly bound van der Waals moecule.

Oxygen. Since the method of determining electronic configurations is clear from the above examples, we shall consider just one more molecule in detail, the oxygen molecule. Filling the orbitals in order of increasing energy the sixteen electrons of O_2 are described by the configuration $1\sigma_g^2 1\sigma_u^2 2\sigma_g^2 2\sigma_u^2 3\sigma_g^2 1\pi_u^4 1\pi_g^2$. The orbital densities are illustrated in Fig.

The molecular orbitals of π symmetry are doubly degenerate and a filled set of π orbitals will contain four electrons. The node in a π_u orbital is in the plane which contains the internuclear axis and is not perpendicular to this axis as is the node in a σ_u orbital. The π_u orbital is therefore bonding. A π_g orbital, on the other hand, is antibonding because it has, in addition to the node in the plane of the bond axis, another at the bond mid-point perpendicular to the axis. The bonding and antibonding characters of the π orbitals have just the opposite relationship to their *g* and *u* dependence as have the σ orbitals.

The $1\sigma_g$ and $1\sigma_u$ orbital densities have, as in the case of Li_2, degenerated into localized atomic distributions with the characteristics of 1*s* core densities and are nonbonding. The valence electrons of O_2 are contained in the remaining orbitals, a feature reflected in the extent to which their density distributions are delocalized over the entire molecule. Aside from the inner nodes encircling the nuclei, the $2\sigma_g$ and $2\sigma_u$ orbital densities resemble the $1\sigma_g$ and $1\sigma_u$ valence density distributions of H_2 and He_2. A quantitative discussion of the relative binding abilities of the $2\sigma_g$, $3\sigma_g$ and 1π orbital densities is presented in the following section.

One interesting feature of the electronic configuration of O_2 is that its outer orbital is not fully occupied. The two π_g electrons could both occupy one of the π_g orbitals with paired spins or they could be assigned one to each of the π_g orbitals and have parallel spins. Hund's principle applies to molecules as well as to atoms and the configuration with single occupation of both π_g orbitals with parallel spins is thus predicted to be the most stable. This prediction of molecular orbital theory regarding the electronic structure of O_2 has an interesting consequence. The oxygen molecule should be magnetic because of the resultant spin angular momentum possessed by the electrons. The magnetism of O_2 can be demonstrated experimentally in many ways, one of the simplest being the observation that liquid oxygen is attracted to the poles of a strong magnet.

The Relative Binding Ability of Molecular Orbitals

We may determine the relative importance of each orbital density in the overall binding of the nuclei in a molecule through a comparison of the forces which the various molecular orbital charge distributions exert on the nuclei. In molecular orbital theory, the total charge density is given by the sum of the orbital charge densities. Thus the total force exerted on the nuclei by the electronic charge distribution will be equal to the sum of the forces exerted by the charge density in each of the molecular orbitals. It is of interest to compare the effectiveness of each orbital charge density in binding the nuclei with some standard case in which they all exhibit the same ability. The limiting forms of the molecular orbitals for the case of the separated atoms have this desired property. In addition, the properties of the separated atoms form a useful basis for the discussion of any molecular property from the point of view of determining the changes which have been brought about by the formation of the chemical bond.

Suppose we label the two nuclei of a homonuclear diatomic molecule as *A* and *B* and consider the forces exerted on the *A* nucleus by the pair of electrons in a molecular orbital when the orbital has assumed its limiting form for the separated atoms. At this limit, one electron correlates with an atomic orbital on nucleus A and the other with an identical orbital on nucleus *B*. The discussion of the forces exerted on the nuclei by such a limiting charge distribution is similar to the discussion given previously in Section 6 for the case of two separated hydrogen atoms. The charge density in the orbital on the A nucleus will not exert a force on that nucleus since an undistorted atomic orbital is centrosymmetric with respect to its nucleus. The charge density of the electron which correlates with the *B* nucleus will exert a force on the *A* nucleus equivalent to that obtained by concentrating the charge density to a point at the position of the *B* nucleus. The electron which correlates with the *B* nucleus will screen one of the nuclear charges of *B* from the *A* nucleus. Thus the force exerted on one of the nuclei by the pair of electrons in a molecular orbital for the limiting state of the separated atoms is equivalent to that obtained by placing one negative charge at the position of the second nucleus. Of the pair of electrons in a given homonuclear molecular orbital, only one is effective in binding either nucleus in the limit of the separated atoms.

If there are a total of N electrons in the molecule, there will be $N/2$ occupied molecular orbitals since each molecular orbital contains a pair of electrons. Therefore, a total of $N/2$ electrons will correlate with each nucleus. The molecule dissociates into two neutral atoms each with a nuclear charge $Z = N/2$. Thus the $N/2$ electrons which correlate with each nucleus will exactly cancel the nuclear charge of both nuclei; the final force on the nuclei will be zero.

The limiting force exerted on the A nucleus by the pair of electrons in a molecular orbital is $(Z_A e^2/R^2)(-1)$, that is, *the force is equivalent to placing one negative charge at the position of the B nucleus*. Thus we may express the total limiting force on nucleus A as the product of $(Z_A e^2/R^2)$ with the difference between the number of positive charges on the B nucleus (Z_B) and *the number of electronic charges which are effective in exerting a force on the A nucleus*, ($N/2$):

$$F_A = \frac{Z_A e^2}{R^2}\left(Z_B - (N/2)\right)$$

The quantity $N/2$ is the *charge equivalent* of the electronic force, the number of charges which when placed at the position of one nucleus exerts the same force on the second nucleus as does the actual charge distribution. The zero force between the separated atoms may be viewed as a result of each electron screening one nuclear charge on one nucleus from the nuclear charge of the other atom.

As the atoms approach one another to form a chemical bond, the atomic distributions on each atom become increasingly distorted and charge density is transferred to the binding region between the nuclei. There is a net force of attraction on the nuclei. We may again express the electronic force on the A nucleus in terms of its charge equivalent by multiplying the electronic force of attraction by $R^2/Z_A e^2$. Because of the distortion of the atomic orbital densities and the formation of molecular orbitals concentrating charge density in the region between the nuclei, the charge density of more than just one electron in each molecular orbital is effective in binding the nuclei. Thus at intermediate R values the charge equivalent of the electronic force exceeds its limiting value of $N/2$ required to screen the nuclear charge and the result is a force of attraction drawing the nuclei together.

When the distance between the nuclei is further decreased to its equilibrium value the force on the nuclei is again equal to zero. At this point, as when R equals infinity, the charge equivalent of the electronic force equals the nuclear charge. However, the state of electrostatic equilibrium in the molecule does not correspond to the charge density in each molecular orbital effectively screening one nuclear charge as it did in the separated atoms. Instead the charge equivalent of the density in each molecular orbital may be less than, equal to, or greater than the limiting value of unity observed for the separated atoms.

An orbital which concentrates charge density in the binding region will exert a force on the nuclei with a charge equivalent greater than unity. Such an orbital is called *binding* as it does more than simply screen one unit of positive charge on each nucleus. The charge equivalent of an orbital which concentrates density in the antibinding regions will be less than the separated atom value of unity. Such an orbital is termed *antibinding* as the charge density does not screen one unit of positive charge on each nucleus. When the charge equivalent of the force equals unity, it implies that the orbital charge density plays the same role in the molecule as in the separated atoms, that of screening one nuclear charge on B from nucleus A. An orbital with this property is termed *nonbinding*.

Thus, by comparing the charge equivalent of the force exerted by the density in each molecular orbital with its separated atom value of unity, we may classify the orbitals as binding, antibinding or nonbinding:

Binding orbital	—	charge equivalent > unity
Nonbinding orbital	—	charge equivalent ~ unity
Antibinding orbital	—	charge equivalent < unity

Table. Charge Equivalents of the Orbital Forces in Homonuclear Diatomic Molecules

Molecule	$1\sigma_g$	$1\sigma_u$	$2\sigma_g$	$2\sigma_u$	$1\pi_u$	$3\sigma_g$	$1\pi_g$	Sum = Z_B	R_e (au)
He_2	1.78	-0.42						1.36*	(1.750)*
Li_2	0.70	0.68	1.62					3.00	5.051
Be_2	1.05	1.08	2.00	-0.40				3.68*	(3.500)*
B_2	0.98	0.98	2.32	-0.48	1.20			5.00	3.005
C_2	0.97	0.95	2.25	-0.43	1.13†			6.00	2.348
N_2	1.15	1.08	2.67	-0.47	1.21†	0.15		7.00	2.068
O_2	1.23	1.14	2.94	-0.52	1.30†	0.18	0.43	8.00	2.282
F_2	1.24	1.12	2.45	-0.16	1.24†	0.52	0.67†	9.00	2.680

He_2 and *Be_2* form only weakly bound van der Waals molecules for relatively large internuclear seperations. The values of *R* quoted for these molecules are the internuclear distances used in the calculation of the charge equivalents listed in the table.

†All of the values are quoted for double occupation of the orbitals for comparative purposes. The values marked by †are to be doubled to obtain the total electronic force as they refer to filled π orbitals.

The charge equivalents of the orbital forces for some homonuclear diatomic molecules are given in Table 8-1. Except for *He_2* and *Be_2* the sum of the charge equivalents equals the nuclear charge in each case as required for electrostatic equilibrium and the formation of a stable molecule. The charge equivalents of the orbital forces provide a quantitative measure of the role each orbital density plays in the binding of the nuclei in the molecule.

The $1\sigma_g$ orbital in He_2 is binding. Of the two electronic charges in the $1\sigma_g$ orbital, 1.78 of them are effective in binding the nuclei when $R = 1.75\ a_0$ as opposed to the one electronic charge which exerts a force when $R = \infty$. The $1\sigma_u$ charge density, however, is strongly antibinding. The transfer of charge density to the antibinding regions in the formation of the $1\sigma_u$ orbital in He_2 is so great that the charge equivalent is negative in sign. The antibinding nature of this orbital is very evident in the form of its charge distribution. Not only does the charge density in this orbital no longer screen a positive charge on one nucleus from the other, it actually exerts a repulsive force on the nuclei, one which pulls the nuclei further apart from one another. The total eletronic force exerted on a nucleus in He_2 at $R = 1.75\ a_0$ is equivalent to placing (1.78 - 0.42) = 1.36 negative charges at the position of the second nucleus. Since the nuclear charge on helium is 2.00, a total of (2.00 - 1.36) = 0.64 positive charges on the second nucleus are left unscreened by the charge density. The net force on the nuclei is thus a repulsive one.

The $1\sigma_g$ and $1\sigma_u$ molecular orbitals are inner shell orbitals in the remaining molecules, Li_2 to F_2. An idealized inner shell molecular orbital has a charge equivalent of unity, the same as the separated atom value. Each electron should be localized in an atomic-like distribution and screen one nuclear charge. This is illustrated by the $1\sigma_g$ and $1\sigma_u$ charge density maps for the O_2 molecule. The charge equivalents of the $1\sigma_g$ and $1\sigma_u$ orbital densities for Li_2 are significantly less than unity. While these orbitals are not as contracted around the nuclei in Li_2 as they are in O_2 (the nuclear charge for lithium is three compared to eight for oxygen), they are still atomic-like with no effective overlap between the two centres. The charge equivalents are less than the screening value of unity because each of the atomic-like distributions is polarized into the antibinding region and exerts an antibinding force on the nucleus on which it is centred. The charge equivalents for the $1\sigma_g$ and $1\sigma_u$ density distributions in the remaining molecules are close to unity indicating that they are essentially nonbinding inner shell orbitals. The slight binding character of the $1\sigma_g$ charge density in O_2 and F_2 is the result of small inward polarizations of the atomic-like distributions.

The $2\sigma_g$ molecular charge density is binding in every molecule. A comparison of the charge equivalents shows that the $2\sigma_g$ charge density is the most binding of all the molecular orbitals in this series of molecules. The charge equivalent of the force exerted by the $2\sigma_g$ density in O_2 is almost three times greater than it is for the separated oxygen atoms. This is a result of the large amount of charge density accumulated in the binding region by this orbital.

The $2\sigma_u$ orbital is uniformly strongly antibinding. The extreme concentration of charge density in the antibinding regions observed for the $2\sigma_u$ orbital is typified by the $2\sigma_u$ density plot for O_2. It is obvious that the density in this orbital, as that in the $1\sigma_u$ orbital of He_2 will pull the nuclei away from one another rather than bind them together. Notice that Be_2 is analogous to He_2 except that the $2\sigma_g$ and $2\sigma_u$ orbitals rather than the $1\sigma_g$and $1\sigma_u$ orbital densities are involved. In Be_2 the $1\sigma_g$ and $1\sigma_u$ densities are nonbinding and together simply screen two nuclear charges on each atom. The $2\sigma_g$ density exerts a binding force equivalent to one electronic charge in excess of the simple screening effect. The $2\sigma_u$ orbital density, however, leaves a single nuclear charge unscreened which cancels the net attractive force of the $2\sigma_g$ density and in addition exerts an antibinding force equivalent to increasing the nuclear charge by 0.40 units. The beryllium molecule is therefore unstable at this value of R.

The $1\pi_u$ orbital density is binding in each case, but only weakly so. The charge density of a π_u molecular orbital is concentrated around the internuclear axis rather than along it as in a σ_g molecular orbital. Consequently the $1\pi_u$ density distributions exert only weak binding forces on the nuclei. In fact, the inner shell $1\sigma_g$ charge density in F_2 exerts as large a binding force on the nuclei as does a pair of electrons in the$1\pi_u$ orbital.

The charge equivalent of the $3\sigma_g$ orbital density is less than unity in the three cases where it is occupied. Thus it is an antibinding orbital even though it is of σ_g symmetry. The charge density contours for this orbital in O_2 show that charge density is accumulated in the region between the nuclei as expected for an orbital of σ_g symmetry. However, the $3\sigma_g$ orbital correlates with a $2p\sigma$ atomic orbital on each nucleus. The strong participation of the $2p\sigma$ orbitals in the molecular orbital is evidenced by the node at each nucleus and by the concentration of charge density on both sides of each nucleus. The concentration of charge in the antibinding regions nullifies the binding effect arising from the accumulation of charge density in the region between the nuclei. The net result is an attractive force considerably less than that required to screen one positive charge on each nucleus.

The $1\pi_g$ orbital density is only weakly antibinding just as the $1\pi_u$ density is only weakly binding. The formation of the $1\pi_g$ orbital results in the removal of charge density from the binding region, not from along the internuclear axis but instead from regions around the axis. Notice that unlike the $2\sigma_u$ orbital densities, the $1\pi_g$ charge density is antibinding only in the sense that it does not screen its share of nuclear charge, not because it exerts a force which draws the nuclei apart.

Molecular Orbitals for Heteronuclear Molecules

The molecular orbitals which describe the motion of a single electron in a molecule containing two unequal nuclear charges will not exhibit the g and u symmetry properties of the homonuclear diatomic case. The molecular orbitals in the heteronuclear case will in general be concentrated more around one nucleus than the other. The orbitals may still be classified as σ, π, δ, etc. because the molecular axis is still an axis of symmetry.

In simple numerical calculations the molecular orbitals are sometimes approximated by the sum and difference of single atomic orbitals on each centre, their limiting form. The molecular orbital is said to be approximated mathematically by a linear combination of atomic orbitals and the technique is known as the LCAO-MO method. It must be understood that the LCAO-MO method using a limited number of atomic orbitals provides only an approximation to the true molecular orbital. The concept of a molecular orbital is completely independent of the additional concept of approximating it in terms of atomic orbitals, except for the case of the separated atoms. However, by using a large number of atomic orbitals centred on each nucleus in the construction of a single molecular orbital sufficient mathematical flexibility can be achieved to approximate the exact form of the molecular orbital very closely.

While the LCAO approximation using a limited number of atomic orbitals is in general a poor one for quantitative purposes, it does provide a useful guide for the prediction of the qualitative features of the molecular orbital. There are two simple conditions which must be met if atomic orbitals on different centres are to interact significantly and form a *molecular orbital which is delocalized over the whole molecule*. Both atomic orbitals must have approximately the same orbital energy and they must possess the same symmetry characteristics with respect to the internuclear axis. We shall consider the molecular orbitals in *LiH*, *CH* and *HF* to illustrate how molecular orbital theory describes the bonding in heteronuclear molecules, and to see how well the forms of the orbitals in these molecules can be rationalized in terms of the symmetry and energy criteria set out above.

The 1*s* and 2*s* atomic orbitals and the 2*p* orbital which is directed along the bond axis are all left unchanged by a rotation about the symmetry axis. They may therefore form molecular orbitals of σ symmetry in the diatomic hydride molecules. The 2*p* orbitals which are perpendicular to the bond axis will be of π symmetry and may form molecular orbitals with this same symmetry. The energies and symmetry properties of the relevant atomic orbitals and the electronic configurations of the atoms and molecules are given in Table.

Table. Atomic Orbital Energies and Symmetry Properties

Energy (au)					**Symmetry**
	H	Li	C	F	
1*s*	-0.5	-2.48	-11.33	-26.38	σ
2*s*		-0.20	-0.71	-1.57	σ
2*p*			-0.43	-0.73	σ and π
Atomic Configurations				Molecular Configurations	
Li	$1s^2 2s^1$			LiH	$1\sigma^2 2\sigma^2$
C	$1s^2 2s^2 2p^2$			CH	$1\sigma^2 2\sigma^2 3\sigma^2 1\pi^1$
F	$1s^2 2s^2 2p^5$			HF	$1\sigma^2 2\sigma^2 3\sigma^2 1\pi^4$

Density diagrams of the molecular orbitals for the *LiH*, *CH*, and *HF* molecules are illustrated in Fig.

The 1*s* orbital energies of *Li*, *C* and *F* all lie well below that of the *H* 1*s* orbital. The charge densities of these inner shell orbitals are tightly bound to their respective nuclei. They should not, therefore, be much affected by the field of the proton or interact significantly with the *H* 1*s* orbital. The molecular orbital of lowest energy in these molecules, the 1σ molecular orbital, should be essentially nonbinding and resemble the doubly-occupied 1*s* atomic orbital on *Li*, *C* and *F* respectively. These predictions are borne out by the 1σ orbital density distributions. They consist of nearly spherical contours centred on the *Li*, *C* and *F* nuclei, the radius of the outer contour being less than the bond length in each case. The forces exerted on the proton by the 1σ charge distributions are equivalent to placing two negative charges at the position of the heavy nucleus in each case. The charge density in the 1σ molecular orbital simply screens two of the nuclear charges on the heavy nucleus from the proton. This same screening effect is obtained for the $1s^2$ charge distribution when the molecules dissociate into atoms. Thus the 1*s* atomic orbitals of *Li*, *C* and *F* are not much affected by the formation of the molecule and the 1σ charge density is nonbinding as far as the proton is concerned. The 1σ atomic-like distributions are slightly polarized. In *LiH* the 1σ density is polarized away from the proton to a significant extent while in *CH* and *HF* it is slightly polarized towards the proton. Thus the 1σ charge density exerts an antibinding force on the *Li* nucleus and a small binding force on the *C* and *F* nuclei.

The energies of the 2*s* atomic orbitals decrease (the electron is more tightly bound) from *Li* to *F* as expected on the basis of the increase in the effective nuclear charge from *Li* to *F*. The 2*s* orbital on *Li* is large and diffuse and will overlap extensively with the 1*s* orbital on *H*. However, the 2*s* electron on *Li* is considerably less tightly bound than is the 1*s* electron on *H*. Thus the charge density of the 2σ molecular orbital in *LiH* will be localized in the region of the proton corresponding to the transfer of the 2*s* electron on *Li* to the region of lower potential energy offered by the 1*s* orbital on *H*. This is approximately correct by the almost complete concentration of the charge density in the region of the proton in the 2σ orbital density map for *LiH*. The small amount of density which does remain around the *Li* nucleus is polarized away from the proton. The 1σ and 2σ densities are polarized in a direction counter to the direction of charge transfer as required in ionic binding. The inwardly polarized accumulation of 2σ charge density centred on the proton binds both nuclei.

The 1σ molecular orbital in *LiH* is to a good approximation a polarized doubly-occupied 1*s* orbital on *Li*, and the 2σ molecular orbital is, to a somewhat poorer approximation, a doubly-occupied and polarized 1*s* orbital on *H*. Our previous discussion of the bonding in *LiH* indicated that the binding is ionic, corresponding to the description $Li^+(1s^2)H^-(1s^2)$. The molecular orbital description of an *ionic bond* is similar in that the molecular orbitals in the ionic extreme are localized in the regions of the individual nuclei, rather than being delocalized over both nuclei as they are for a covalent bond.

The matching of the 2*s* orbital energy with the *H* 1*s* orbital energy is closer in the case of *C* than it is for *Li*. Correspondingly, the 2σ charge density in *CH* is delocalized over both nuclei rather than concentrated in the region of just one nucleus as it is in the *LiH* molecule. There is a considerable buildup of charge density in the binding region which is shared by both nuclei. The 2σ charge density exerts a large binding force on both the *H* and *C* nuclei. This is the molecular orbital description of an interaction
which is essentially covalent in character.

The 2*s* orbital energy of *F* is considerably lower than that of the *H* 1*s* orbital. The 2σ orbital charge density in *HF*, therefore, approximately resembles a localized 2*s* orbital on *F*. It is strongly polarized and distorted by the proton, but the amount of charge transferred to the region between the nuclei is not as large as in *CH*. The 2σ orbital in *HF* plays a less important role in binding the proton than it does in *CH*.

The 3σ molecular orbital in *CH* and *HF* will result primarily from the overlap of the 2*p*σ orbital on *C* and *F*, with the 1*s* orbital on *H*. The 2*p*-like character of the 3σ molecular orbital in both

CH and *HF* is evident in the density diagrams. In *CH* the 1*s* orbital of *H* interacts strongly with both the 2*s* and 2$p\sigma$ orbitals on *C*. In terms of the forces exerted on the nuclei, the 2σ charge density is strongly binding for both *C* and *H*, while the 3σ charge density is only very weakly binding for *H* and is actually antibinding for the *C*. This antibinding effect is a result of the large accumulation of charge density in the region behind the *C* nucleus.

In *HF*, the *H* 1*s* orbital interacts only slightly with the 2*s* orbital on *F*, but it interacts very strongly with the 2$p\sigma$ orbital in the formation of the 3σ molecular orbital. The 3σ charge density exerts a large binding force on the proton. Thus the proton is bound primarily by the 2σ charge density in *CH* and by the 3σ charge density in *HF*. The 3σ charge density in *HF* is primarily centred on the *F* nucleus and roughly resembles a 2$p\sigma$ orbital. Although no density contours are actually centred on the proton, the proton is embedded well within the orbital density distribution. This is a molecular orbital description of a highly polar bond.

The 3σ orbital charge density exerts a force on the *F* and *C* nuclei in a direction away from the proton. The molecular orbitals which involve $p\sigma$ orbitals are characteristically strongly polarized in a direction away from the bond in the region of the nucleus on which the *p* orbital is centred. Compare, for example, the 3σ orbitals of *CH* and HF with the $3\sigma_g$ molecular orbital of the homonuclear diatomic molecules.

When the *C* and *H* atoms are widely separated, we can consider the carbon atom to have one 2*p* electron in the 2$p\sigma$ orbital which lies along the bond axis, and the second 2*p* electron in one of the 2$p\pi$ orbitals which are perpendicular to the bond. The F atom has five 2*p* electrons and of these one may be placed in the 2$p\sigma$ orbital; the remaining four 2*p* electrons will then completely occupy the 2$p\pi$ orbitals. The singly-occupied 2$p\sigma$ orbitals on *F* and *C* eventually interact with the singly-occupied 1*s* orbital on H to form the doubly-occupied 3σ molecular orbital in HF and CH. The remaining 2*p* electrons, those of π symmetry, will occupy the 1π molecular orbital. The H atom does not possess an orbital of π symmetry in its valence shell and the vacant 2$p\pi$ orbital on H is too high in energy (-0.125 au) to interact significantly with the

$2p\pi$ orbitals on C and F. Thus the 1π molecular orbital is atomic-like, centred on the F and C nuclei and is essentially nonbinding. The 1π molecular orbital resembles a $2p\pi$ atomic orbital in each case, but one which is polarized in the direction of the proton.

The 1π orbitals of *CH* and *HF* illustrate an interesting and general result. In the formation of a bond between different atoms, the charge density in the σ orbitals is transferred from the least to the most electronegative atom. However, the charge density of π symmetry, if any is present, is invariably transferred, or at least polarized, in the opposite direction, towards the least electronegative atom. Although the amount of charge density transferred is less in the formation of the π orbitals than in the σ orbitals, one effect increases with the other. Thus the polarization is more pronounced in *HF* than in *CH*.

The three examples considered above demonstrate the essential points of a molecular orbital description of the complete range of chemical bonding. In the ionic extreme of *LiH* the charge density of the bonding molecular orbital is localized around the proton. In *CH* the valence charge density is more evenly shared by both nuclei and the bond is covalent. The motions of the electrons in *HF* are governed largely by the potential field of the *F* nucleus. This is evidenced by the appearance of the molecular orbital charge distributions. The proton is, however, encompassed by the valence charge density and the result is a polar bond.

MOLECULAR ORBITALS FOR POLYATOMIC MOLECULES

The concept of a molecular orbital is readily extended to provide a description of the electronic structure of a polyatomic molecule. Indeed molecular orbital theory forms the basis for most of the quantitative theoretical investigations of the properties of large molecules.

In general a molecular orbital in a polyatomic system extends over all the nuclei in a molecule and it is essential, if we are to understand and predict the spatial properties of the orbitals, that we make use of the symmetry properties possessed by the nuclear framework. An analysis of the molecular orbitals for the water molecule provides a good introduction to the way in which the symmetry of a molecule determines the forms of the molecular orbitals in a polyatomic system.

There are three symmetry operations which transform the nuclear framework of the water molecule into itself and hence leave the nuclear potential field in which the electrons move unchanged.

For each symmetry operation there is a corresponding symmetry *element*. The symmetry elements for the water molecule are a two-fold axis of rotation C_2 and two planes of symmetry σ_1 and σ_2. A rotation of 180° about the C_2 axis leaves the oxygen nucleus unchanged and interchanges the two hydrogen nuclei. A reflection through the plane labelled σ_1 leaves all the nuclear positions unchanged while a reflection through σ_2 interchanges the two protons. The symmetry operations associated with the three symmetry elements either leave the nuclear positions unchanged or interchange symmetrically equivalent (and hence indistinguishable) nuclei. Every molecular orbital for the water molecule must, under the same symmetry operations, be left unchanged or undergo a change in sign.

Similarly we may use the symmetry transformation properties of the atomic orbitals on oxygen and hydrogen together with their relative orbital energy values to determine the primary atomic components of each molecular orbital in a simple LCAO approximation to the exact molecular orbitals. Only atomic orbitals which transform in the same way under the symmetry operations may be combined to form a molecular orbital of a given symmetry. The symmetry transformation properties of the atomic orbitals on oxygen and hydrogen are given in Table. A value of +1 or -1 opposite a given orbital in the table indicates that the orbital is unchanged or changed in sign respectively by a particular symmetry operation.

Table. Symmetry Properties and Orbital Energies for the Water Molecule

Atomic Orbitals on Oxygen	Symmetry Behaviour Under			Symmetry Classification	1	Orbital Energy (au)
	σ_2	σ_1	σ_2			
1*s*	+1	+1	+1	a_1		–20.669
2s	+1	+1	+1	a_1		–1.244
$2p_z$	+1	+1	+1	a_1	}	–0.632
$2p_x$	–1	+1	–1	b_2	}	
$2p_y$	–1	–1	+1	b_1	}	

Atomic Orbitals on Hydrogen					
$(1s_1 - 1s_2)$	+1	+1	+1	a_2	} –0.500
$(1s_1 - 1s_2)$	–1	+1	–1	b_2	}

Molecular Orbital Energies for H_2O (au)

$1a_1$	$2a_1$	$1b_2$	$3a_1$	$1b_1$
–20.565	–1.339	–0.728	–0.595	–0.521

The $1s$, $2s$ and $2p_z$ orbitals of oxygen are symmetric (i.e., unchanged) with respect to all three symmetry operations. They are given the symmetry classification a_1. The $2p_x$ orbital, since it possesses a node in the σ_2 plane (and hence is of different sign on each side of the plane) changes sign when reflected through the σ_2 plane or when rotated by 180° about the C_2 axis. It is classified as a b_2 orbital. The $2p_y$ orbital is antisymmetric with respect to the rotation operator and to a reflection through the σ_1 plane. It is labelled b_1.

The hydrogen $1s$ orbitals when considered separately are neither unchanged nor changed in sign by the rotation operator or by a reflection through the σ_2 plane. Instead both these operations interchange these orbitals. The hydrogen orbitals are said to be symmetrically equivalent and when considered individually they do not reflect the symmetry properties of the molecule. However, the two linear combinations $(1s_1 + 1s_2)$ and $(1s_1 - 1s_2)$ do behave in the required manner. The former is symmetric

under all three operations and is of a_1 symmetry while the latter is antisymmetric with respect to the rotation operator and to a reflection through the plane σ_2 and is of b_2 symmetry.

The molecular orbitals in the water molecule are classified as a_1, b_1 or b_2 orbitals, as determined by their symmetry properties. This labelling of the orbitals is analogous to the use of the σ–π and g-u classification in linear molecules. In addition to the symmetry properties of the atomic orbitals we must consider their relative energies to determine which orbitals will overlap significantly and form delocalized molecular orbitals.

The $1s$ atomic orbital on oxygen possesses a much lower energy than any of the other orbitals of a_1 symmetry and should not interact significantly with them. The molecular orbital of lowest energy in H_2O should therefore correspond to an inner shell 1*s* atomic-like orbital centred on the oxygen. This is the first orbital of a_1 symmetry and it is labelled $1a_1$. Reference to the forms of the charge density contours for the la, molecular orbital substantiates the above remarks regarding the properties of this orbital.

The atomic orbital of next lowest energy in this system is the 2*s* orbital of a_1 symmetry on oxygen. We might anticipate that the extent to which this orbital will overlap with the $(1s_1 + 1s_2)$ combination of orbitals on the hydrogen atoms to form the $2a_1$ molecular orbital will be intermediate between that found for the 2σ molecular orbitals in the diatomic hydrides *CH* and *HF*. The 2σ orbital in CH results from a strong mixing of the 2*s* orbital on carbon and the hydrogen 1*s* orbital. In HF the participation of the hydrogen orbital in the 2σ orbital is greatly reduced, a result of the lower energy of the 2*s* atomic orbital on fluorine as compared to that of the 2*s* orbital on carbon.

Aside from the presence of the second proton, the general form and nodal structure of the $2a_1$ density distribution in the water molecule is remarkably similar to the 2σ distributions in *CH* and HF, and particularly to the latter. The charge density accumulated on the bonded side of the oxygen nucleus in the $2a_1$ orbital is localized near this nucleus as the corresponding charge increase in the 2σ orbital of HF is localized near the fluorine. The charge density of the $2a_1$ molecular orbital accumulated in the region between the three nuclei will exert a force drawing all three nuclei together. The $2a_1$ orbital is a binding orbital.

Although the three 2*p* atomic orbitals are degenerate in the oxygen atom the presence of the two protons results in each 2*p* orbital experiencing a different potential field in the water molecule. The nonequivalence of the 2*p* orbitals in the water molecule is evidenced by all three possessing different symmetry properties. The three 2*p* orbitals will interact to different extents with the protons and their energies will differ.

The $2p_x$ orbital interacts most strongly with the protons and forms an orbital of b_2 symmetry by overlapping with the $(1s_1 - 1s_2)$

combination of $1s$ orbitals on the hydrogens. The charge density contours for the lb_2 orbital indicate that this simple LCAO description accounts for the principal features of this molecular orbital. The $1b_2$ orbital concentrates charge density along each *O-H* bond axis and draws the nuclei together. The charge density of the $1b_2$ orbital binds all three nuclei. In terms of the forces exerted on the nuclei the $2a_1$ and lb_2 molecular orbitals are about equally effective in binding the protons in the water molecule.

The $2p_z$ orbital may also overlap with the hydrogen $1s$ orbitals, the $(1s_1 + 1s_2)\ a_1$ combination, and the result is the $3a_1$ molecular orbital. This orbital is concentrated along the *z*-axis and charge density is accumulated in both the bonded and nonbonded sides of the oxygen nucleus. It exerts a binding force on the protons and an antibinding force on the oxygen nucleus, a behaviour similar to that noted before for the 3σ orbitals in *CH* and *HF*.

The $2p_y$ orbital is not of the correct symmetry to overlap with the hydrogen $1s$ orbitals. To a first approximation the $1b_1$ molecular orbital will be simply *a* $2p_y$ atomic orbital on the oxygen, perpendicular to the plane of the molecule. Reference to Fig indicates that the $1b_1$ orbital does resemble *a* $2p$ atomic orbital on oxygen but one which is polarized into the molecule by the field of the protons. The $1b_1$ molecular orbital of H_2O thus resembles a single component of the 1π molecular orbitals of the diatomic hydrides. The $1b_1$ and the 1π orbitals are essentially nonbinding. They exert a small binding force on the heavy nuclei because of the slight polarization. The force exerted on the protons by the pair of electrons in the $1b_1$ orbital is slightly less than that required to balance the force of repulsion exerted by two of the nuclear charges on the oxygen nucleus. The $1b_1$ and 1π electrons basically do no more than partially screen nuclear charge on the heavy nuclei from the protons.

In summary, the electronic configuration of the water molecule as determined by molecular orbital theory is

$$1a^2{}_1 2a^2{}_1 1b^2{}_2 3a^2{}_1 1b^2{}_1$$

The la_1 orbital is a nonbinding inner shell orbital. The pair of electrons in the la_1 orbital simply screen two of the nuclear charges on the oxygen from the protons. The $2a_1$, $1b_2$ and $3a_1$ orbitals accumulate charge density in the region between the nuclei and the charge densities in these orbitals are responsible for binding the protons in the water molecule. Aside from being polarized by the presence of the protons, the $1b_1$ orbital is a non-interacting $2p_y$ orbital on the oxygen and is essentially nonbinding.

Before closing this introductory account of molecular orbital theory, brief mention should be made of the very particular success which the application of this theory has had in the understanding of the chemistry of a class of organic molecules called conjugated systems. Conjugated molecules are planar organic molecules consisting of a framework of carbon atoms joined in chains or rings by alternating single and double bonds. Some examples are

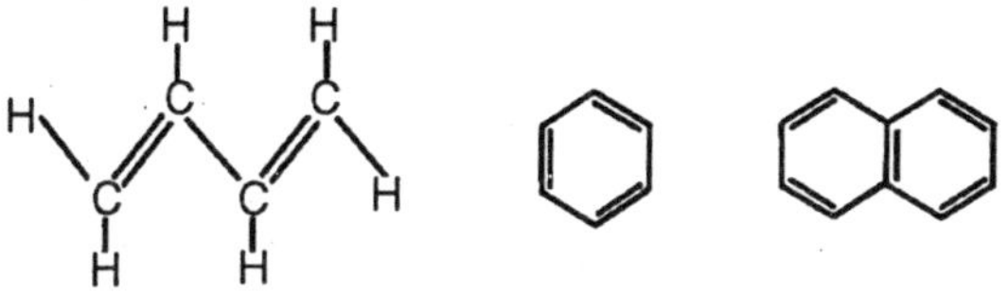

butadiene benzene Napthalene

In the structural formulae for the cyclic molecules, e.g., benzene and naphthalene, it is usual not to label the positions of the carbon and hydrogen atoms by their symbols. A carbon atom joined to just two other carbon atoms is in addition bonded to a hydrogen atom, the *C—H* bond axis being projected out of the ring in the plane of the carbon framework and bisecting the *CCC* bond angle.

The notion of these molecules possessing alternating single and double bonds is a result of our attempt to describe the bonding in terms of conventional chemical structures. In reality all six *C—C* bonds in benzene are identical and the *C—C* bonds in the other two examples possess properties intermediate between those for single and double bonds. In other words, the pairs of electrons forming the second or p bonds are not localized between specific carbon atoms but are delocalized over the whole network of carbon atoms, a situation ideally suited for a molecular orbital description.

We may consider each carbon atom in a conjugated molecule to be sp^2 hybridized and bonded through these hybrid orbitals to three other atoms in the plane. This accounts for the bonding of the hydrogens and for the formation of the singly-bonded carbon network. The electrons forming these bonds are called s electrons. The axis of the remaining 2*p* orbital on each carbon atom is directed perpendicular to the plane of the molecule and contains a single electron, called a p electron. A simple adaptation of molecular orbital theory, called Hückel theory, which takes the s bonds for granted and approximates the molecular orbitals of the p electrons in terms of linear combinations of the 2*p*p atomic orbitals on each carbon atom, provides a remarkably good explanation of the properties of conjugated molecules.

Index

F

H

I

L

M

N